Cysticercosis of the Human Nervous System

Oscar H. Del Brutto • Héctor H. García

Cysticercosis of the Human Nervous System

Springer

Oscar H. Del Brutto
School of Medicine
Universidad Espíritu Santo
Ecuador

Department of Neurological Sciences
Hospital-Clinica Kennedy
Guayaquil
Ecuador

Héctor H. García
Cysticercosis Unit
Instituto Nacional de Ciencias Neurológicas
Lima
Peru

School of Sciences
Universidad Peruana Cayetano Heredia
Lima
Peru

ISBN 978-3-642-39021-0 ISBN 978-3-642-39022-7 (eBook)
DOI 10.1007/978-3-642-39022-7
Springer Heidelberg New York Dordrecht London

Library of Congress Control Number: 2013946240

Printed on acid-free paper

Springer is part of Springer Science+Business Media (www.springer.com)

Preface

Neurocysticercosis—defined as the infection of the central nervous system and its covering by the larval stage of the tapeworm *Taenia solium*—is a major public health challenge for most of the developing world and is increasingly recognized in developed countries due to mass immigration, refugee movements, international travelling, and overseas business affairs. Known since ancient civilizations as a disease of swine and recognized as a human ailment during the Renaissance, it has not been until the past few decades when the introduction of modern neuroimaging techniques and reliable serologic tests allowed the preoperative recognition of neurocysticercosis and increased our knowledge on the many clinical forms of presentation of this parasitic disease. This, together with the availability of potent cysticidal drugs, has changed the prognosis of most affected patients. In addition, well-conducted epidemiologic surveys have changed our views on basic mechanisms of disease acquisition. Now, human cysticercosis must be considered a disease mainly transmitted from person to person, i.e., a healthy person ingesting *T. solium eggs* directly from a Taenia carrier by the fecal-oral route or through non-hygienic handling of food. In this context, the role of infected swine is to perpetuate the infection cycle only (human develop taeniasis after ingesting pork containing cysticerci).

As occurs with most zoonosis, the disease complex of taeniasis/cysticercosis is closely linked to poverty and ignorance, disproportionately affecting underserved populations. Despite the above-mentioned advances in diagnosis and therapy, neurocysticercosis is still the most common cause of acquired epilepsy in the world, with conservative estimates suggesting that 50 million people are infected by cysticerci. Much remains to be learned on some controversial aspects of this highly pleomorphic disease. This book intends to provide a comprehensive—and yet practical—review of basic and clinical aspects of human *T. solium* cysticercosis that will be of value to physicians of different specialties involved in the care of patients affected by this old scourge.

Guayaquil, Ecuador — Oscar H. Del Brutto
Lima, Peru — Héctor H. García

Contents

1 History of Taeniasis and Cysticercosis

The history of parasitic diseases—including taeniasis—is as old as that of humanity (Cox 2004), but it has been only in the past few centuries that enough knowledge has been accumulated to proper characterize the disease complex taeniasis/cysticercosis as a major infection of human beings (Fig. 1.1).

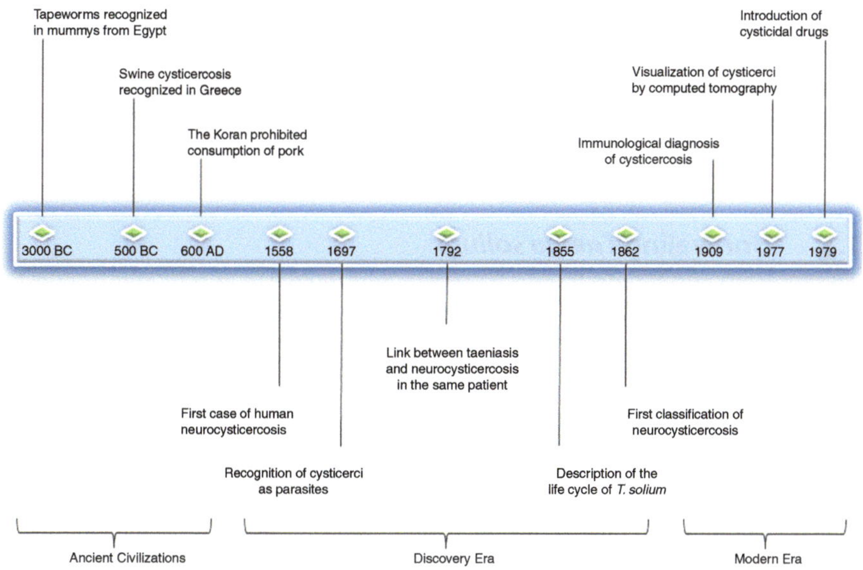

Fig. 1.1 Timeline of major events in the history of *Taenia solium* taeniasis and cysticercosis

O.H. Del Brutto, H.H. García, *Cysticercosis of the Human Nervous System*, DOI 10.1007/978-3-642-39022-7_1, © Springer-Verlag Berlin Heidelberg 2014

1.1 Ancient Knowledge

The Ebers Papyrus (1500 BC) includes descriptions of tapeworms and both *Taenia* sp. eggs, and *Taenia solium* cysticerci have been found in the intestine and stomach of mummies from the ancient Egypt (Bruschi et al. 2006; Le Bailly et al. 2010). Intestinal tapeworms were familiar among physicians of different ancient civilizations, who treated them with pumpkin seeds (*Cucurbita pepo*), a herbal medicine that it is still used nowadays (Li et al. 2012). Also, the occurrence of swine cysticercosis (measly pork) was a common knowledge among the ancient Greeks. Aristophanes (*c*. 448–385 BC) in his comedy "The Knights" (424 BC) has one of the slaves inspecting someone's tongue "to see if he is measled." Aristotle (*c*. 384–322 BC) in his book *History of Animals* described the presence of muscle bladders or cysts in pig muscles that were compared with hailstones. Interestingly, Aristotle recognized that nursing pigs do not suffer from the disease (Nieto 1982). The fact that swine was considered impure in the ancient Greece probably inclined Muhammad (570–632 AD), the Islam founder, to prohibit the consumption of pork when he wrote the Koran.

While the association between taeniasis, human cysticercosis, and epilepsy was far to be even considered in ancient times, there is some indirect evidence suggesting that cysticercus-related epilepsy may have been present in ancient Greece and Rome. The concept that epilepsy starting in adults could be related to a structural disease of the nervous system may be traced back to the Hippocratic treatise "On the Sacred Disease" (Daras et al. 2008), and there is some evidence supporting the cysticercotic origin of the epilepsy that afflicted the Roman dictator Gaius Julius Cesar (100–44 BC), as it started to happen 1 year after one of his visits to Egypt (Bruschi 2011; McLachlan 2010).

1.2 Unraveling *Taenia solium*

While human taeniasis and swine cysticercosis were well known in ancient civilizations, proper identification of the causal agent, and the fact that both conditions are related to the same helminth, was not recognized for centuries. Theophrastus (*c*. 372–287 BC) and Pliny (25–79 AD) have been credited for coining the term *Taenia* (from the Greek, ribbon or band). However, the name *solium* was used centuries later by Arnaldo de Villanueva (*c*. 1300 AD) to indicate that humans usually host a single tapeworm. Indeed, the denomination of *Taenia solium*—meaning in Latin "the lonely worm"—reflects this traditional, though erroneous, view (Del Brutto et al. 1998). There was a long-lasting confusion between the different tapeworms affecting humans, including *T. saginata*, *T. solium*, and *Diphyllobothrium latum*. In 1683, the discovery of the scolex by Tyson provided the first solid element for the separation of these species, and during the second half of the eighteenth century, *T. solium* was finally introduced into the modern zoological classification as a distinct helminth (Nieto 1982).

Despite this knowledge, there were a lot of inaccurate thoughts on the behavior of *T. solium*, like that of Alpheus Myer, an American physician who, as late as 1854, designed and patented a trap for the tapeworm (Fig. 1.2). According to his beliefs,

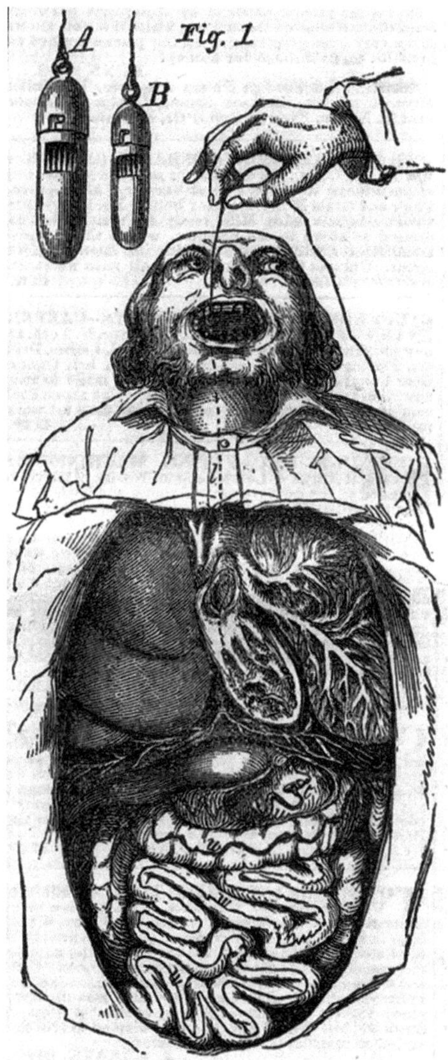

Fig. 1.2 Diagram of the Myers tapeworm trap (From de Souza N. Ingestion/The beast within). *Cabinet Magazine* 2009 (© Cabinet Magazine, with permission)

if the patient is starved for several days, the worm will move up to the stomach looking for food and may be trapped with a device embedded in cheese that has been previously introduced into the patient's stomach and secured with a silk thread that will be used to pull out the parasite through the mouth (de Souza 2009).

1.3 First Descriptions of Human Cysticercosis

Several writers concur that the first recorded case of human neurocysticercosis was probably that described by Rumler in 1558, during the autopsy of a patient with epilepsy who had liquid-filled vesicles adherent to the meninges. In the seventeenth century, Paranolus found similar vesicles in the corpus callosum of a priest who had suffered from seizures, and Wharton found cysts in the muscles of a soldier (Del Brutto et al. 1998; Nieto 1982; Wadia and Singh 2002). However, the correct parasitic nature of these vesicles was not recognized by that time, and it was Malpighi, in 1697, who described the head of the *T. solium* inside them. During those years, Gmelin coined the term *Taenia cellulosae* for the vesicles, and Zeder included them in a new genus, *Cysticercus* (from the Greek, bladder and tail). It was believed that cysticercus constituted a separate species of parasite, and it was classified as *Cysticercus cellulosae* due to its tendency to develop in connective tissue; this incorrect term is still widely used nowadays (Del Brutto et al. 1998).

1.4 Taeniasis and Cysticercosis Were Caused by the Same Parasite

The occurrence of taeniasis and neurocysticercosis in the same person was probably first recognized by the Peruvian physician, journalist, and politician Hipólito Unanue (Fig. 1.3) in 1792, as he wrote in the journal *El Mercurio* the case of a soldier with taeniasis who died following a generalized seizure (Deza 1987). Some years later, German pathologists confirmed the morphological similarities between the head of the adult *T. solium* and the scolex of cysticercus, and Küchenmeister (1855) demonstrated that oral consumption of cysticercus from pork resulted in human intestinal taeniasis by feeding convicted men with a soup containing cysticerci obtained from a recently slaughtered pig. At autopsy, Küchenmeister found "a small *Taenia* that was tightly attached with its proboscis to a piece of duodenal mucosa," as well as other nine taenias, one of them with the complete crown of 22 hooklets in two rows typical of the rostellum of *T. solium*. The definition of the life cycle of *T. solium* was completed during the second half of the nineteenth century by experiments in Belgium and Germany, demonstrating that when *Taenia* eggs obtained from proglottids passed by infected humans were fed to pigs, the animals develop cysticercosis; failed attempts to infect other animals demonstrated that the main hosts of *T. solium* are humans and swine (Del Brutto et al. 1998; Nieto 1982). These findings were confirmed later by the work of Yoshino (1933a, b, c) who infected himself with *T. solium* cysticerci to study the life cycle of the cestode.

1.5 Pioneer Clinical Descriptions

Relevant clinical advances on the knowledge of neurocysticercosis during the second half of the nineteenth century and the turn of the twentieth century included the description of meningeal cysticercosis ("traubenhydatiden") by Virchow, the first classification of the clinical syndromes of cysticercosis by Griesinger (Table 1.1), the description of cysticercotic angiitis by Askanazy, and the recognition of transient episodes of loss of consciousness in patients with fourth ventricle cysticercosis by Bruns (Arellano-Sánchez et al. 1985; Nieto 1982; Trelles and Trelles 1978).

Fol. 108.

Núm 117.

MERCURIO PERUANO

DEL DIA 16 DE FEBRERO DE 1792.

115.

OBSERVACION MÉDICA.

VENTURA Medez Soldado del Reximiento Real de Lima entró el dia 8 del presente mes de Febrero en el Hospital de S. Andres acometido de pasmo. Era este de la clase de aquellos que el Pueblo llama *pasmo de arco*, y los Médicos *opistótonos*. Fué socorrido con el uso de los oleosos, así interno como externo, y arrojó de resulta de él una lombriz de aquellas que nombran los Médicos *tenias*, la que tenia cinco varas y sesma. Independente de este retazo unido, salieron otras porciones menores, y todas juntas formaban cerca de siete varas. El paciente murió el dia 10, tercero del insulto. En su cadáver podria el cuchillo Anatómico haber manifestado los desórdenes que aquella formidable lombriz causó en sus órganos, y si tomaron de aquí orígen las contracciones de los músculos; pero no habiéndose concluido la habilitacion del Anfiteatro-Anatómico, faltan los recursos prontos para la indagacion de unas causas, sin cuyo conocimiento jamas podrá adelantarse la Medicina Peruana. Por fortuna vivimos baxo de un Gobernador esclarecido que va á perfeccionar esta obra interesante, y dexar en ella á la posteridad un nuevo monumento de sus luces y beneficencia.

Fig. 1.3 (*Left*) Postage stamp honoring the Peruvian physician and journalist Hipólito Unanue (1755–1833), probably the first person who recognized the coexistence of *Taenia solium* taeniasis and neurocysticercosis in the same person, and (*Right*) original publication of the journal *El Mercurio*, dating from February 1792, where it was described such association (downloaded from http://www.cervantesvirtual.com/obra-visor/mercurio-peruano–18/html/027f5ec8-82b2-11df-acc7-002185ce6064_129.htm and http://www.cervantesvirtual.com/obra-visor/mercurio-peruano–18/html/027f5ec8-82b2-11df-acc7-002185ce6064_136.htm, last accessed April 15, 2013, documents in the open domain)

Table 1.1 Griesinger's 1862 classification of neurocysticercosis

Asymptomatic cases
Epileptic patients without mental disease
Epileptics with associated psychiatric disease
Cases of chronic mental disease alone
Patients with signs and symptoms of diffuse cerebral irritation

In Germany, the book *Handbuch der Neurologie* included a chapter completely devoted to *Cysticercosis cellulosae* quoting references from central Europe, where this pathology was well recognized at the turn of the twentieth century (Henneberg 1912). Such chapter collected all the knowledge on human cysticercosis by that time, and many of the concepts written there are still valid nowadays.

Other European relevant descriptions of the disease by those years were those of Davaine (1877), author of one of the first textbooks of parasitology, who wrote extensively on neurocysticercosis; Volovatz (1902) who reported 414 cases of human cysticercosis—including 149 with neurocysticercosis—in his thesis "Ladrerie du cysticercose chez l'homme;" and Guccione (1919) and Schmite (1928) who published the firsts books on neurocysticercosis. In Asia, neurocysticercosis was first recognized in 1888 during the autopsy of a patient from a lunatic asylum in Madras (Wadia 1995), and in Latin America, several case reports came from different countries by the end of the nineteenth century and the turn of the twentieth century (Trelles and Trelles 1978). The first US cases of neurocysticercosis occurring in immigrants were recognized during the first half of the twentieth century (Dolgopol and Neustaedter 1935), far before the epidemic bout of neurocysticercosis in Mexican immigrants that occurred in the Southwestern US during the 1970s and 1980s (Richards et al. 1985).

1.6 Lessons from History

Two natural epidemiological scenarios improved significantly our knowledge on the mechanisms of disease transmission of human cysticercosis and allowed us to better understand the consequences of migration of humans or swine from disease-endemic to non-endemic areas.

1.6.1 The British Endemia

A bout of neurocysticercosis in the United Kingdom—elegantly called by Wadia and Singh (2002) as the "British Military Connection"—was observed from 1930 to 1950, after the massive return of British troops stationed in India. It was MacArthur (Fig. 1.4) who first noticed the occurrence of cerebral cysticercosis among soldiers with epilepsy returning from their duties abroad (MacArthur 1934). MacArthur (1935) also noted the high prevalence of asymptomatic cases of neurocysticercosis, as he wrote *For every case of cysticercosis immediately diagnosticable there are a large number which will defy diagnosis for years*. Other British physicians reinforced these findings by reporting a total of 450 British patients with cysticercosis (Dixon and Hargreaves 1944; Dixon and Lipscomb 1961; Dixon and Smithers 1934). It could also be recorded the onset of neurological symptoms according to their date of return to a cysticercosis-free country by that time, and it was noticed that most patients had their initial seizure between 2 and 5 years after return, suggesting a long asymptomatic stage. Most important, some patients were family members of these soldiers but had never been in India, suggesting locally transmitted disease from a soldier household contact with neurocysticercosis that was also a *T. solium* carrier (Dixon and Lipscomb 1961). British doctors also noticed that treatment of Taeniasis may lead to exacerbations of symptoms of cysticercosis in some cases (Dixon and Smithers 1934).

Fig. 1.4 Lieutenant General Sir William Porter MacArthur (1884–1964) (© National Portrait Gallery, London, UK, with permission)

1.6.2 The Epidemic of Burns in West New Guinea

Despite the swine-breading culture of the inhabitants of West New Guinea, the region was free of swine cysticercosis until 1972, where Ekari people of the Enaratoli region received a number of infected pigs as a gift from the Indonesian government of Java. By the next year, the presence of cysticerci was noticed in the flesh of local pigs slaughtered in this region, and soon thereafter, intestinal taeniasis was diagnosed in humans. By 1974, an epidemic of burns was observed among Ekari people (Subianto et al. 1978). Such an epidemic was related to seizures causing people to fall into the bonfires that they used to warm-up the huts during the cool mountain nights. Many of these epileptic patients also had subcutaneous nodules that were pathologically confirmed as cysticercus, and one of them (a girl dying after the seizure episode) had thousands of parasites into the brain parenchyma (Gajdusek 1978). Some years later, cysticercosis remained endemic in that area and started to spread to neighboring regions (Bendin and Catford 1983), and even nowadays, the disease is still prevalent (Salim et al. 2009), showing the difficulties that exist to eradicate the parasite once it has been established.

1.7 Modern Times: Advances in Diagnosis and Treatment

The introduction of the complement fixation test signaled the start of a new age, since the diagnosis of human cysticercosis could be—for the very first time—established without the need of histological demonstration of the parasite (Teive et al. 2006). The complement fixation test was first developed by Weinberg (1909), who demonstrated antibodies against cysticercus in the serum of infected pigs. Thereafter, Moses (1911) obtained positive results with the use of this test in the cerebrospinal fluid of patients with neurocysticercosis, and Lange (1940) and Nieto (1956) settled the role of the complement fixation test in the diagnosis of the meningeal form of the disease. More recent advances in neuroradiology, particularly computed tomography, allowed the correct localization of intracranial cysticerci and improved the diagnosis of the disease (Bentson et al. 1977; Carbajal et al. 1977). Finally, the advent of specific therapy for human cysticercosis (Robles and Chavarría 1979) opened the more recent chapter in the history of this ancient scourge. With the development of potent cysticidal drugs, improved diagnostic methods, and effective vaccines against swine cysticercosis, it is expected that the work of hundreds of scientists will conclude with the eradication of human cysticercosis.

References

Arellano-Sánchez J, Aruffo C, Escobar A (1985) Cisticercosis del IV ventrículo y síndrome de Bruns. Rev Fac Med UNAM 28:11–19

Bendin JJ, Catford JC (1983) Epidemic of burns in New Guinea due to cerebral cysticercosis. Lancet 1:922

Bentson JR, Wilson GH, Helmer E, Winter J (1977) Computed tomography in intracranial cysticercosis. J Comput Assist Tomogr 1:464–471

Bruschi F (2011) Was Julius Caesar's epilepsy due to neurocysticercosis? Trends Parasitol 27: 373–374

Bruschi F, Masetti M, Locci MT, Ciranni R, Fornaciari G (2006) Cysticercosis in an Egyptian mummy of the late Ptolemaic period. Am J Trop Med Hyg 74:598–599

Carbajal JR, Palacios E, Azar-Kia B, Churchill R (1977) Radiology of cysticercosis of the central nervous system including computed tomography. Radiology 125:127–131

Cox FEG (2004) History of human parasitic diseases. Infect Dis Clin North Am 18:171–188

Daras MD, Bladin PF, Eadie MJ, Millett D (2008) Epilepsy: historical perspectives. In: Engel J Jr, Pedley TA (eds) Epilepsy. A comprehensive approach, 2nd edn. Lippincott Williams & Wilkins, Philadelphia

Davaine C (1877) Traité des entozoaires et des maladies vermineuses de l'homme et des animaux domestiques, Deuxieme edition. J-B Bailliére et Fils, Paris

de Souza N (2009) Ingestion/the beast within. Cabinet Magazine 2009; 34. www.cabinetmagazine.org/issues/34/desouza.php, downloaded on 07 July 2012

Del Brutto OH, Sotelo J, Román GC (1998) Neurocysticercosis. A clinical handbook. Sweets & Zeitlinger Publishers, Lisse

Deza L (1987) Hipólito Unanue y la neurocisticercosis. Rev Neuro-Psiquiatría (Perú) 50:77–82

Dixon HBF, Hargreaves WH (1944) Cysticercosis (*Taenia solium*). A further ten years' clinical study covering 284 cases. Q J Med 13:107–121

Dixon HBF, Lipscomb FM (1961) Cysticercosis: an analysis and follow-up of 450 cases, Medical research council special report series no 299. Her Majesty's Stationary Office, London

Dixon HBF, Smithers DW (1934) Epilepsy in cysticercosis (*Taenia solium*): a study of seventy-one cases. QJM 3:603–616

Dolgopol VB, Neustaedter M (1935) Meningo-encephalitis caused by Cysticercus cellulosae. Arch Neurol Psychiatry 33:132–147

Gajdusek DC (1978) Introduction of *Taenia solium* into West New Guinea with a note on an epidemic of burns from cysticercus epilepsy in the Ekari people of the Wissel Lakes area. P N G Med J 21:329–342

Guccione A (1919) La cisticercosis del sistema nervioso centrale umano. Libraria, Milano

Henneberg R (1912) Die tierischen Parasiten des Zentralnervensystems. I. Des *Cysticercus cellulosae*. In: Lewandowsky M (ed) Handbuch der Neurologie, vol 3, Spezielle Neurologie II. Verlag von Julius Springer, Berlin

Küchenmeister F (1855) Offenes sendschreiben an die k.k. Gessellschaft der Aertze zu Wein. Experimenteller Nachweis, dass *Cysticercus cellulosae* innerhab des menschlichen Darmkanales sich in *Taenia solium* umwandelt. Wiener Medizinische Wochenscriff 5:1–4

Lange O (1940) Sindrome liquorico na cisticercose encéfalo meningeia. Arch Neuropsiquiat 2:322–326

Le Bailly M, Mouze S, Chaves da Rocha G, Heim J-L, Lichtenberg R, Dunand F, Bouchet F (2010) Identification of *Taenia* sp. in a mummy from a Christian necropolis in El-Deir, oasis of Kharga, ancient Egypt. J Parasitol 96:213–215

Li T, Ito A, Chen X, Long C, Okamoto M, Raoul F, Girardoux P, Yanagida T, Nakao M, Sako Y, Xiao N, Craig PS (2012) Usefulness of pumpkin seeds combined with areca nut extract in community-based treatment of human taeniasis in northwest Sichuan Province, China. Acta Trop 124(2):152–157

MacArthur WP (1934) Cysticercosis as seen in the British Army with special reference to the production of epilepsy. Trans R Soc Trop Med Hyg 27:343–363

MacArthur WP (1935) Cysticercosis of the brain. Br Med J 2:1229

McLachlan RS (2010) Julius Caesar's late onset epilepsy: a case of historic proportions. Can J Neurol Sci 37:557–561

Moses A (1911) Dos métodos biológicos de diagnóstico nas cisticercozes. Mem Inst Oswaldo Cruz 3:322–326

Nieto D (1956) Cysticercosis of the nervous system. Diagnosis by means of the spinal fluid complement fixation test. Neurology 6:725–738

Nieto D (1982) Historical notes on cysticercosis. In: Flisser A, Willms K, Laclete JP, Larralde C, Ridaura C, Beltrán F (eds) Cysticercosis: present state of knowledge and perspectives. Academic, New York

Richards FO Jr, Ruiz-Tiben E, Schantz PM, Sorvillo FJ (1985) Cysticercosis in Los Angeles county. JAMA 254:3444–3448

Robles C, Chavarría M (1979) Presentación de un caso clínico de cisticercosis cerebral tratado médicamente con un nuevo fármaco: praziquantel. Salud Púb Méx 21:603–617

Salim L, Ang A, Handali S, Tsang VCW (2009) Seroepidemiologic study of cysticercosis-taeniasis in four central highland districts of Papua, Indonesia. Am J Trop Med Hyg 80:384–388

Schmite P (1928) La cysticercose du systéme nerveux. Legrand, Paris

Subianto DB, Tumada LR, Margono SS (1978) Burns and epileptic fits associated with cysticercosis in mountain people of the Irian Jaya. Trop Geogr Med 30:275–278

Teive HAG, de Almeida SM, Werneck LC (2006) The Brazilian contribution to the study of neurocysticercosis. Moses and Lange's role in cerebrospinal fluid diagnosis. Arq Neuropsiquiatr 64:534–537

Trelles JO, Trelles L (1978) Cysticercosis of the nervous system. In: Vinken PJ, Bruyn GW (eds) Handbook of clinical neurology, vol 3. North Holland, Amsterdam

Volovatz A (1902) Ladrerie du cysticercose chez l'homme. Thesis Med, Paris

Wadia NH (1995) Neurocysticercosis in Asia. Presented at the informal consultation of taeniasis/cysticercosis, Pan American Health Organization, Brasilia, Brazil, 23–25 Aug 1995

Wadia NH, Singh G (2002) *Taenia solium*: a historical note. In: Singh G, Prabhakhar S (eds) *Taenia solium* cysticercosis. From basic to clinical science. CABI Publishing, Oxon

Weinberg M (1909) Recherche des anticorps spécifiques dans la distomatose et la cysticercose. C R Soc Biol (Paris) 66:219–221
Yoshino K (1933a) Studies on the post-embryonal development of *Taenia solium*. Part I. On the hatching of the eggs of *Taenia solium*. J Med Assoc Formos 32:139–141
Yoshino K (1933b) Studies on the post-embryonal development of *Taenia solium*. Part II. On the migratory course of the oncosphera of *Taenia solium* within the intermediate host. J Med Assoc Formos 32:155–158
Yoshino K (1933c) Studies on the post-embryonal development of *Taenia solium*. Part III. On the development of Cysticercus cellulosae within the definite intermediate host. J Med Assoc Formos 32:166–169

Taenia solium: Biological Characteristics and Life Cycle

2

Taenia solium is one of several species of cestodes (tapeworms) that can infect humans (Flisser 1994; Pawlowski 2002). It belongs to the phylum Platyhelminthes, class Cestoidea, order Cyclophyllidea, family Taeniidae. In general terms, tapeworms are complex organisms having complex life cycles that require at least two hosts for their completion. In the case of *Taenia solium*, humans are the most important definitive hosts, whereas both pigs and humans are the main intermediate hosts. Other animals may act as definitive or intermediate hosts of *Taenia solium* but their participation is clinically irrelevant.

2.1 Developmental Stages of *Taenia solium*

Different developmental stages of *Taenia solium* must be recognized, including the adult tapeworm, the egg, the embryo (oncosphere), and the larva (cysticercus). The adult *T. solium* has a head (scolex) that consists of four suckers and a rostellum equipped with a double crown of hooks, a narrow neck, and a large body composed by hundreds of proglottids (Fig. 2.1). The scolex of *T. solium* is unique in its characteristics, allowing its easy differentiation from that of *T. saginata* which is not armed with hooks. The body of the parasite is called the strobila and may be several meters long. Each proglottid of *T. solium* is equipped with a rudimentary nervous system, male and female reproductive organs, and a complex absorption and excretory system (Cárdenas-Ramirez et al. 1982). Proglottids located proximal to the scolex are immature and lack sexual organs. More sexually mature segments are located in the center of the strobila. Those mature proglottids copulate (with itself or with neighboring proglottids) and begin producing eggs. Eggs-containing proglottids are located at the distal end of the strobila. They are referred to as "gravid proglottids" and are ready to be detached from the rest of the tapeworm by a process of apolysis. Gravid proglottids contain thousands of fertile eggs which are accumulated within a long and profusely branched central uterus (Fig. 2.2).

O.H. Del Brutto, H.H. García, *Cysticercosis of the Human Nervous System*,
DOI 10.1007/978-3-642-39022-7_2, © Springer-Verlag Berlin Heidelberg 2014

Fig. 2.1 Proglottid of *Taenia solium* (Image in the public domain, provider CDC/Dr. Mae Melvin)

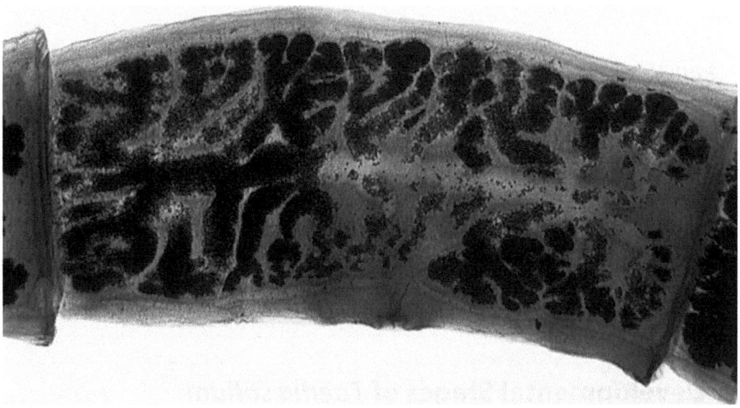

Fig. 2.2 Scolex of *Taenia solium* (Image in the public domain, provider CDC/Dr. Mae Melvin)

The adult *Taenia solium* inhabits the small intestine of humans, where it is attached to the intestinal wall by its suckers and hooks. The parasite lives for years in the hostile intestinal environment, as it has been adapted to survive to variations in pH and to the effects of digestive enzymes. Two or three times per week, a few gravid proglottids are detached from the distal end of the body of the worm and are passed with feces. Each proglottid liberates up to 50,000 eggs which are very resistant to adverse environments; eggs can remain viable for months in water, soil, and vegetation, particularly in humid and warm environments (Aluja et al. 1987; Flisser et al. 1986).

Taenia solium eggs consist of an inner part called the oncosphere and a surrounding coat or embryophore (Fig. 2.3). The oncosphere is the parasite itself in its embryonal stage, and the embryophore is a rigid structure that protects the embryo from external environmental conditions. The latter is formed by radially distributed blocks of a keratin-like protein attached to each other by a cementing substance (Flisser 1994; Laclette et al. 1982). Once in the intestinal tract of their intermediate host (either humans or pigs), the embryophore is disrupted by enzymatic digestion. The liberated oncosphere is a solid globular organism composed of muscular, excretory, and nervous cells, as well as six embryonic hooklets—hence the term "hexacanth embryo"—and a pair of glands that are useful for migration (Pawlowski 2002). Oncospheres cross the intestinal wall and enter the bloodstream, from where they are

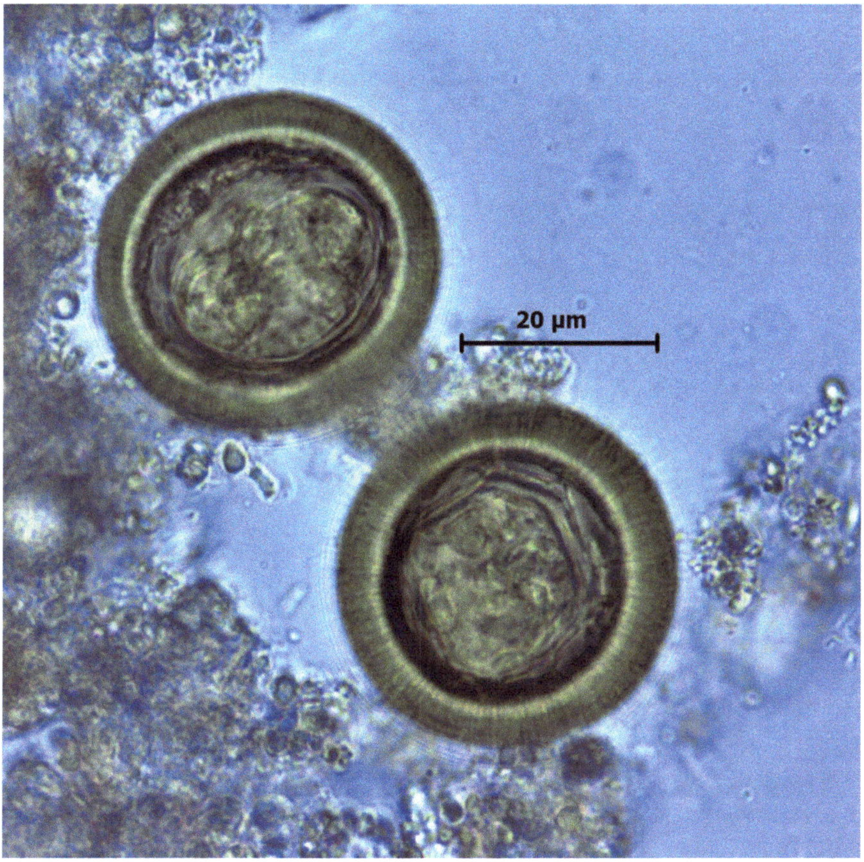

Fig. 2.3 Microscopic appearance of *Taenia solium* eggs (Courtesy of Juan Jimenez, MD, The Cysticercosis Working Group in Peru)

carried to the tissues of the intermediate host where they rapidly evolve into metacestodes—the so-called postoncospheral stage—and then into larvae (cysticercus).

Taenia solium cysticercus is a vesicle that contains an invaginated scolex (Fig. 2.4). The vesicular wall is a membranous structure with festooned appearance composed of an outer eosinophilic layer called the cuticular mantle, a middle cellular layer with pseudoepithelial structure, and an inner layer formed by circular muscle and reticular fibers (Davis and Kornfeld 1991; Pittella 1997). The cuticular mantle is covered by microtriches which, in turn, are coated by a carbohydrate glycocalyx that increases the absorptive surface of the cysticercus. The glycocalyx is the most antigenic anatomical structure of the cysticercus. A network of excretory and neural structures is also seen inside the vesicular wall. The structure of the vesicular wall may be seen as a tegument through which adult taenias and cysticerci fulfill their metabolic and nutritional needs by absorption and diffusion (Lumsden et al. 1982; Thomas et al. 1989). The vesicular fluid is mainly composed of water but also contains calcium, glycoproteins, cholinesterase, and coproporphyrin; such composition confers the vesicular fluid fluorescent properties as well as antigenicity (Cervantes et al. 1986; Martínez-Zedillo et al. 1982).

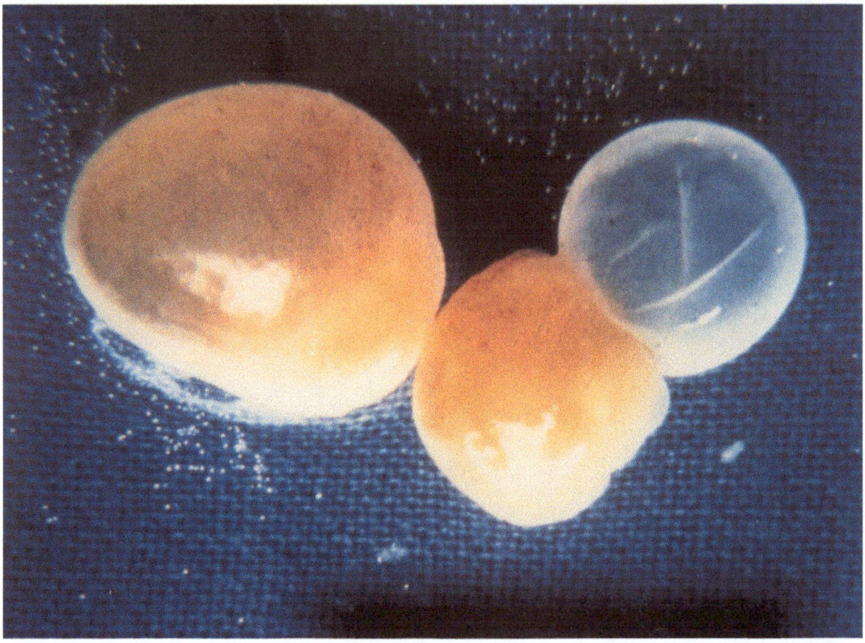

Fig. 2.4 Macroscopic appearance of *Taenia solium* cysticerci. Showing vesicle and invaginated scolex (From Del Brutto et al. (1998), with permission)

The scolex located inside the vesicle has a similar structure than the adult *Taenia solium*, including a head armed with suckers and hooks, an elongated neck, and a rudimentary body (Fig. 2.5). Hooks are anhistous structures arranged in a double crown, with 22–28 small hooks in the outer crown, and a similar number of slightly larger hooks in the inner crown. As the larva grows, neck and body bent in a spiral within the vesicle originating the so-called spiral canal which leads from the bladder wall to the scolex. The structure of the tegument of the neck and the body of the larvae is similar to the vesicular wall, except that the cuticular mantle is thicker, and calcareous corpuscles are found in the reticular layer (Fig. 2.6).

Much of our knowledge on the different developmental stages of *Taenia solium* is derived from the experiments of Yoshino (1933a, b, c) who infected himself with *Taenia solium* cysticerci to have a constant stock of eggs to experimentally infect pigs in order to study the stages of development of embryos and metacestodes in the natural intermediate host. Yoshino observed developing metacestodes measuring 0.3 mm in the host's tissues 6 days after oral infection. By 12 days metacestodes became cystic, and by 20–30 days, a rudimentary scolex appeared. The cysts continued to grow, measuring from 3 to 8 mm at 40–50 days. By then, cysticerci were completely formed but continued to grow for up to 1 year after infection.

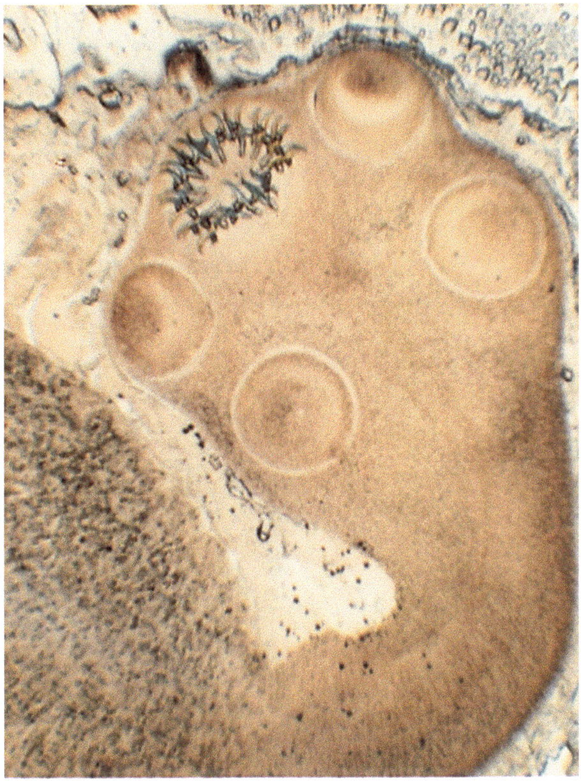

Fig. 2.5 Cysticercus of *Taenia solium* showing scolex with four suckers and double crown of hooks (From Del Brutto et al. (1998), with permission)

2.2 Life Cycle of *Taenia solium*

As noted, the life cycle of *Taenia solium* usually involves humans and pigs (Fig. 2.7). In the normal cycle of transmission, humans are definitive hosts harboring the adult parasite in the small intestine. Gravid proglottids, detached from the distal end of the adult worm, contain infective eggs that are passed with feces of taenia carriers.

In places with improper disposal of human feces and poor husbandry, pigs have access to human feces, and in the event that those feces contain *Taenia solium* eggs, pigs get infected and become intermediate hosts in the life cycle of this cestode (Corwin et al. 1986; Gonzalez et al. 2002; Jayashi et al. 2012; Prasad et al. 2011). This is most commonly observed in rural areas of endemic countries, where all the conditions favoring the completion of the life cycle of *Taenia solium* are combined, including illiteracy, poverty, open-air defecation, warm climate, allowance of free-roaming pigs, and consumption of improperly cooked pork under poor hygienic conditions.

After being ingested by pigs (the natural intermediate host of *Taenia solium*), eggs cross the intestinal wall, enter the bloodstream, and mature into oncospheres

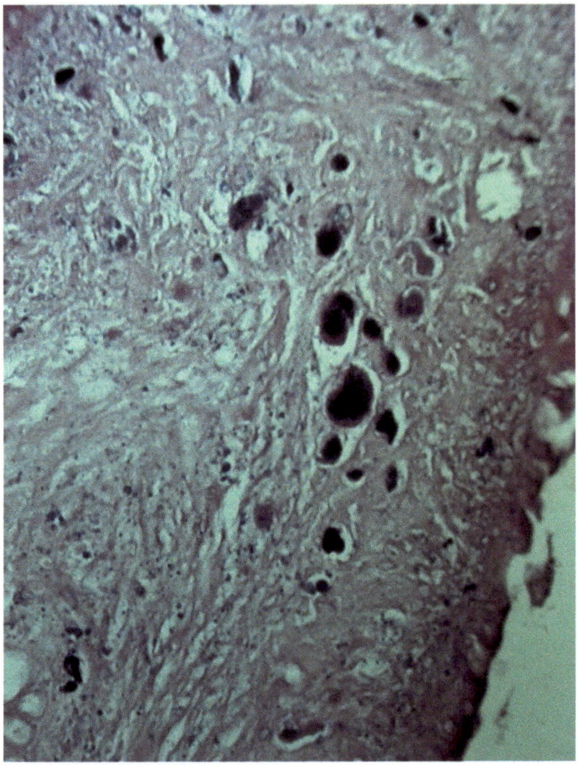

Fig. 2.6 Calcareous corpuscles in the membrane of *Taenia solium* cysticercus

and then into metacestodes which lodge in the host tissues (particularly striated muscles). Once lodged, metacestodes evolve into cysticerci. Pigs can be infected by thousands of cysticerci, giving rise to the so-called measly meat (Fig. 2.8). Human consumption of improperly cooked measly pork results in release of cysticerci in the small intestine. The common practice of eating fried pork in developing countries—at least in rural areas Latin America—is a major source of human infection with *Taenia solium* eggs, since large blocks of pork (measuring about 3 cm^3) are fried for just a few minutes in a large boil and this does not allow the destruction of eggs that are located inside the block (Fig. 2.9).

Once in the human stomach, eggs are liberated from their coat by the action of bile and digestive enzymes. Then, scolices evaginate and attach to the intestinal wall. After the scolex is attached, the rudimentary body begins to grow, forming proglottids. Proglottids begin to multiply and will become mature enough to be excreted in feces approximately 4 months after infection, thus completing the life cycle of the parasite (Aluja et al. 1987).

The adult *Taenia solium* may live for years in the small intestine. While not impossible, it is unusual for humans to be infected with more than one *Taenia solium* at a time, as the first attached tapeworm induces some type of tissue immunity that prevents further attachment of other evaginated scolices to the bowel wall (Flisser et al. 1979). Taeniasis in humans is a benign condition, mainly producing abdominal discomfort and peripheral eosinophilia (Botero et al. 1993).

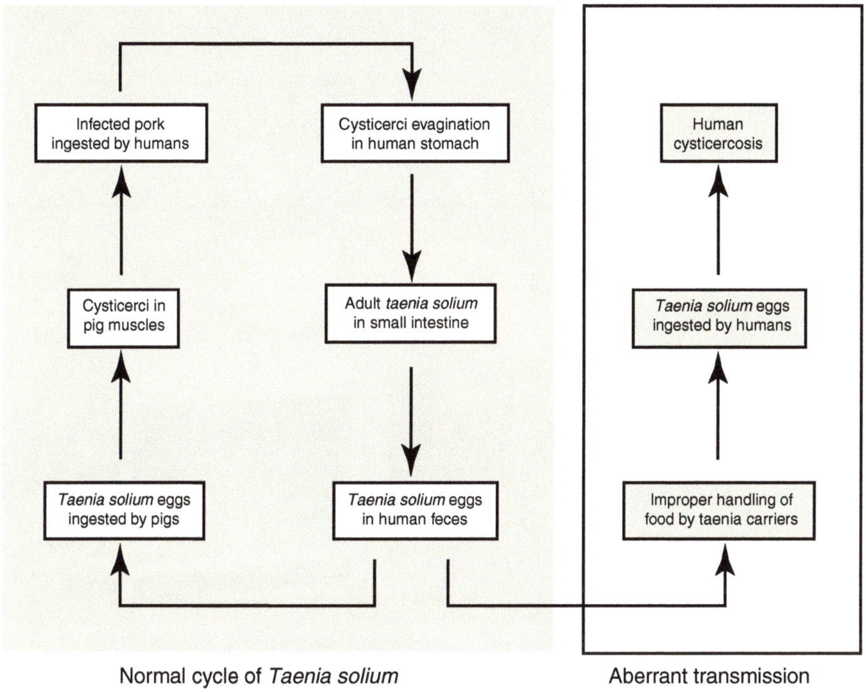

Fig. 2.7 Diagram of major steps in the life cycle of *Taenia solium* (Reproduced with permission from: Del Brutto (2012b))

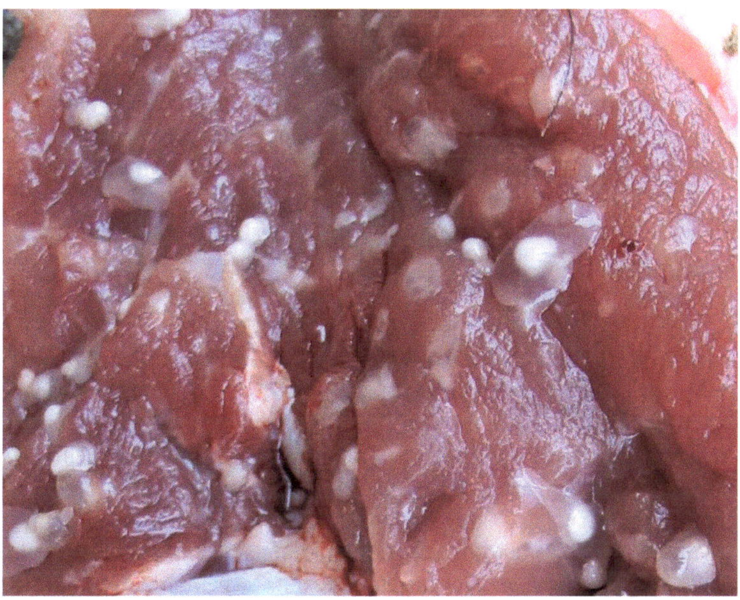

Fig. 2.8 Pork muscles infected with hundreds of cysticerci, giving rise to the so-called measly pork (Courtesy of the Cysticercosis Working Group in Perú)

Fig. 2.9 Consumption of undercooked pork under poor sanitary conditions is a threat to thousands of people living in rural villages of developing countries (Reproduced with permission from: Del Brutto (2012b))

Humans can also act as intermediate hosts for *Taenia solium* after ingesting its eggs. Under these circumstances, human cysticercosis develops (Tay-Zavala 1983). The mechanisms by which eggs cross the intestinal wall and lodge in human tissues are the same as those described in pigs. Unfortunately, the central nervous system is one of the main target organs of cysticerci in humans; this infection produces a complex disease, neurocysticercosis.

The main source from which humans acquire cysticercosis is by contact with a taenia carrier, which may infect the individual directly from the fecal-oral route or through non-hygienic manipulation of food. Unfortunately, anecdotal data—mainly coming from Mexican studies performed some decades ago—suggested that the most common form of human acquisition of the disease was through the ingestion of vegetables or fruits grown in plantations fertilized with manure containing human feces contaminated with *Taenia solium* eggs (Escobar 1983). Indeed, strawberries from a region called "El Bajío" were seen as the main culprit for the high prevalence of human cysticercosis in Mexico City and surrounding areas. Several years later, it was demonstrated that such strawberries were free of taenia eggs, demystifying their role in the transmission of the disease to humans (Spindola Felix et al. 1996). Other studies have also shown negative results when attempting to recover taenia eggs from nature, even in highly endemic areas (Diaz et al. 1992;

Keilbach et al. 1989). More recent studies, showing clusters of cysticercosis patients around taenia carriers, changed definitively the old concept that environmental dispersion is the main source of human contamination with taenia eggs, confirming the major role of person-to-person transmission of the disease (Lescano et al. 2009).

The airborne transmission of *Taenia solium* eggs through flying vectors or internal autoinfection by the regurgitation of proglottids into the stomach of patients with taeniasis has also been proposed as possible mechanisms of infection (Tay-Zavala 1983); however, these are less likely to occur and have not been convincingly demonstrated. Other proposed anecdotal mechanisms of infection included the transplacental transmission of larvae (Escobar and Nieto 1972) and infection from drinking beverages contaminated with *Taenia solium* eggs as part of herbal medicines provided by some traditional healers in Africa (Bourke and Petana 1994; Heinz and Macnab 1965).

To conclude, human cysticercosis must be seen as a disease mostly transmitted from person-to-person, and the role of infected pigs is to perpetuate the life cycle of *Taenia solium*. In this way, the occurrence of autochthonous cases of neurocysticercosis in non-endemic regions where swine husbandry is adequate or inexistent is understandable. Asymptomatic taenia carriers, emigrated from endemic regions, may transmit the disease through non-hygienic handling of food or by direct contact. The predominance of person-to-person transmission of cysticercosis also explains the high prevalence of neurocysticercosis in the Indian subcontinent, where most of the population is vegetarian. As will be discussed later on this book, the most common form of disease expression in India, as well as in non-endemic regions, is the so-called single cysticercus granuloma, a benign form of neurocysticercosis that is highly suggestive of sporadic contact with a taenia carrier (Del Brutto 2012a; Del Brutto et al. 2012; García et al. 2010; Singh et al. 2010). Taenia carriers may also infect themselves through the fecal-oral route. Due to repetitive autoinfections, those patients develop mechanisms of immunosuppression that allow the implantation of hundreds of parasites in the nervous system without triggering a severe inflammatory reaction or major symptoms (García and Del Brutto 1999; Gilman et al. 2000).

References

Aluja A, Escobar A, Escobedo F, Flisser A, Laclete P, Larralde C, Madrazo I, Velásquez V, Willms K (1987) Cisticercosis. Una recopilación actualizada de los conocimientos básicos para el manejo y control de la cisticercosis causada por *Taenia solium*. Biblioteca de la Salud, Mexico City

Botero D, Tanowitz HB, Weiss LM, Wittner M (1993) Taeniasis and cysticercosis. Infect Dis Clin North Am 7:683–697

Bourke GJ, Petana WB (1994) Human taenia cysticercosis: a bizarre mode of transmission. Trans R Soc Trop Med Hyg 88:680

Cárdenas-Ramirez L, Zaragoza AM, González-del-Pliego M (1982) Neural and excretory structures of *Cysticercus cellulosae*. In: Flisser A, Willms K, Laclete JP, Larralde C, Ridaura C, Beltrán F (eds) Cysticercosis: present state of knowledge and perspectives. Academic, New York

Cervantes M, González-Angulo A, Márquez-Monter H (1986) Anatomía bioquímica del *Cysticercus cellulosae*. I. Estudio histoquímico y análisis por energía dispersiva de rayos X: ácidos mucopolisacáridos, glucógeno, grasa, AND, calcio y fierro. Patología (Méx) 24: 209–218

Corwin RM, DiMarco NK, McDowell AE, Pratt SE (1986) Internal parasites. In: Leman AD, Straw B, Glock RD, Mengeling WL, Penny RHC, Scholl E (eds) Diseases of swine. Iowa State University Press, Ames

Davis LE, Kornfeld M (1991) Neurocysticercosis: neurologic, pathogenic, diagnostic and therapeutic agents. Eur Neurol 31:229–240

Del Brutto OH (2012a) Neurocysticercosis among international travelers to disease-endemic areas. J Travel Med 19:112–117

Del Brutto OH (2012b) Neurocysticercosis. Continuum (Minneap Minn). 18:1392–1416

Del Brutto OH, Sotelo J, Del Brutto OH (eds) (1998) Neurocysticercosis. A clinical handbook. Swets & Zeitlinger Publishers, Lisse

Del Brutto OH, Nash TE, García HH (2012) Cysticerci-related single parenchymal brain enhancing lesions in non-endemic countries. J Neurol Sci 319:32–36

Diaz F, García HH, Gilman RH, Gonzales AE, Castro M, Tsang VC, Pilcher JB, Vasquez LE, Lescano M, Carcamo C, Madico G, Miranda E (1992) Epidemiology of taeniasis and cysticercosis in a Peruvian village. Am J Epidemiol 135:875–882

Escobar A (1983) The pathology of neurocysticercosis. In: Palacios E, Rodriguez-Carbajal J, Taveras JM (eds) Cysticercosis of the central nervous system. Charles C. Thomas Publisher, Springfield

Escobar A, Nieto D (1972) Parasitic diseases. In: Minckler J (ed) Pathology of the nervous system, vol 3. McGraw-Hill Book Company, New York

Flisser A (1994) Taeniasis and cysticercosis due to *Taenia solium*. In: Sun T (ed) Progress in clinical parasitology. CRC Press, Boca Raton

Flisser A, Pérez-Montfort R, Larralde C (1979) The immunology of human and animal cysticercosis: a review. Bull World Health Organ 57:839–856

Flisser A, Avidan Y, Laiter S, Mintz D, Ongay H (1986) Efecto de agentes físicos y químicos sobre la viabilidad del cisticerco de la *Taenia solium*. Salud Púb Méx 28:551–555

García HH, Del Brutto OH (1999) Heavy nonencephalitic cerebral cysticercosis in tapeworm carriers. Neurology 53:1582–1584

García HH, Gonzalez AE, Rodriguez S, Tsang VC, Pretell EJ, Gonzales I, Gilman RH (2010) Neurocysticercosis. Unraveling the nature of the single cysticercal granuloma. Neurology 75:654–658

Gilman RH, Del Brutto OH, García HH, Martinez M (2000) Prevalence of taeniasis among patients with neurocysticercosis is related to severity of infection. Neurology 55:1062

Gonzalez AE, Wilkins PP, Lopez T (2002) Porcine cysticercosis. In: Singh G, Prabhakhar S (eds) *Taenia solium* cysticercosis. From basic to clinical science. CABI Publishing, Oxon

Heinz HJ, Macnab GM (1965) Cysticercosis in the Bantu of Southern Africa. S Afr J Med Sci 30:19–31

Jayashi CM, Arroyo G, Lightowlers MW, García HH, Rodriguez S, Gonzalez AE (2012) Seroprevalence and risk factors for *Taenia solium* cysticercosis in rural pigs of Northern Peru. PLoS Negl Trop Dis 6(7):e1733

Keilbach NM, De Aluja AS, Sarti E (1989) A programme to control taeniasis-cysticercosis (*Taenia solium*): experience in a Mexican village. Acta Leiden 57:181–189

Laclette JP, Ornelas V, Merchant MT, Willms K (1982) Ultrastructure of the surrounding envelopes of *Taenia solium* eggs. In: Flisser A, Willms K, Laclete JP, Larralde C, Ridaura C, Beltrán F (eds) Cysticercosis: present state of knowledge and perspectives. Academic, New York

Lescano AG, García HH, Gilman RH, Gavidia CM, Tsang VC, Rodriguez S, Moulton LH, Villaran MV, Montano SM, Gonzalez AE (2009) *Taenia solium* cysticercosis hotspots surrounding tapeworm carriers: clustering of human seroprevalence but not on seizures. PLoS Negl Trop Dis 3(1):e371

Lumsden RD, Voge M, Sogandares-Bernal F (1982) The metacestode tegument: fine structure, development, topochemistry, and interactions with the host. In: Flisser A, Willms K, Laclete JP, Larralde C, Ridaura C, Beltrán F (eds) Cysticercosis: present state of knowledge and perspectives. Academic, New York

Martínez-Zedillo G, González-Barranco D, Pérez-González M, González-Angulo A (1982) Cholinesterases of *Cysticercus cellulosae*. In: Flisser A, Willms K, Laclete JP, Larralde C, Ridaura C, Beltrán F (eds) Cysticercosis: present state of knowledge and perspectives. Academic, New York

Pawlowski ZS (2002) *Taenia solium*: basic biology and transmission. In: Singh G, Prabhakhar S (eds) *Taenia solium* cysticercosis. From basic to clinical science. CABI Publishing, Oxon

Pittella JEH (1997) Neurocysticercosis. Brain Pathol 7:681–693

Prasad KN, Verma A, Srivastava S, Gupta RK, Pandey CM, Paliwal VK (2011) An epidemiological study of asymptomatic neurocysticercosis in a pig farming community in northern India. Trans R Soc Trop Med Hyg 105:531–536

Singh G, Rajshekhar V, Murthy JM, Prabhakar S, Modi M, Khandelwal N, García HH (2010) A diagnostic and therapeutic scheme for a solitary cysticercus granuloma. Neurology 75: 2236–2245

Spindola Felix N, Rojas Wasta Vino G, de Haro Arteaga I, Cabrera Bravo M, Salazar Schettino PM (1996) Parasite search in strawberries from Irapuato, Guanajuato and Zamora, Michoacan (Mexico). Arch Med Res 27:229–231

Tay-Zavala J (1983) Etiology of cysticercosis. In: Palacios E, Rodriguez-Carbajal J, Taveras JM (eds) Cysticercosis of the central nervous system. Charles C. Thomas Publisher, Springfield

Thomas JA, Knoth R, Schwechheimer K, Volk B (1989) Disseminated human neurocysticercosis. Acta Neuropathol 78:594–604

Yoshino K (1933a) Studies on the post-embryonal development of *Taenia solium*. Part I. On the hatching of the eggs of *Taenia solium*. J Med Assoc Formos 32:139–141

Yoshino K (1933b) Studies on the post-embryonal development of *Taenia solium*. Part II. On the migratory course of the oncosphera of *Taenia solium* within the intermediate host. J Med Assoc Formos 32:155–158

Yoshino K (1933c) Studies on the post-embryonal development of *Taenia solium*. Part III. On the development of cysticercus cellulosae within the definite intermediate host. J Med Assoc Formos 32:166–169

3 Neuropathology of Cysticercosis

Almost every organ of the human economy may be affected by cysticerci, but with few exceptions, significant disease is mostly observed in patients with neurocysticercosis. The latter is a highly pleomorphic disease, defined as the infection of the central nervous system and its coverings by the larval stage of the tapeworm *T. solium*. The clinical pleomorphism of neurocysticercosis is mainly related to both the myriad of pathological changes that parasites may cause within the nervous system and the individual differences that exist in the severity of the host's immune response against cysticerci which, in turn, causes a great variety of lesions in the adjacent cerebral tissue (Fleury et al. 2010; Mahanty and García 2010).

3.1 Microscopic Identification of the Parasite

As previously described, cysticercus is a vesicle consisting of two main parts, the vesicular wall and the scolex. When the whole vesicle is available for pathological examination, it is useful to open the vesicle until the larva is exposed. Then, the larva is placed between two glass slides which are pressed firmly until the scolex flattens; subsequent microscopic examination usually reveals the characteristic double crown of hooks and the four suckers of *Taenia solium* (Fig. 3.1). Examination of serial sections of the vesicle also allows the proper differentiation of *Taenia solium* cysticercus from the larvae of other cestodes (Tay-Zavala 1983). This permits the visualization of the spiral canal and the scolex, the latter appearing as a compact structure having in its interior several anhistous, cornified, and semitransparent structures corresponding to the hooks (Fig. 3.2).

In some cysticerci—particularly those located in the subarachnoid space—the scolex cannot be identified. Instead, these parasites are mainly composed of several membranes attached to each other (Bickerstaff et al. 1952; Escobar 1983). It has been shown that membranes begin to proliferate after the scolex degenerates, showing changes in their histochemical composition, including the presence of large amounts of hydrophilic lipids (Valkounova et al. 1992).

O.H. Del Brutto, H.H. García, *Cysticercosis of the Human Nervous System*,
DOI 10.1007/978-3-642-39022-7_3,

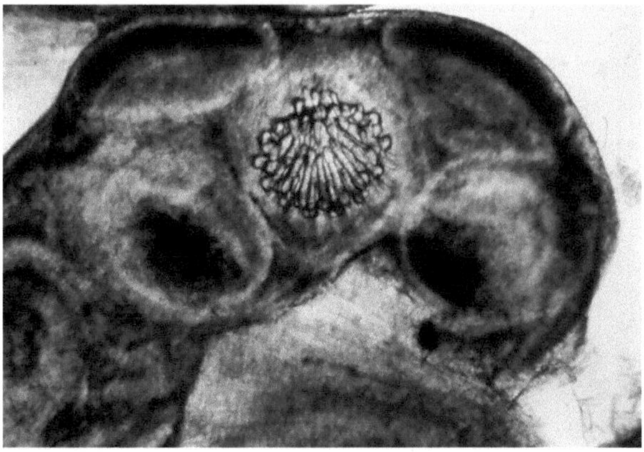

Fig. 3.1 Rostellum of *Taenia solium*, fresh specimen seen under light microscopy (From Del Brutto et al. (1998), with permission)

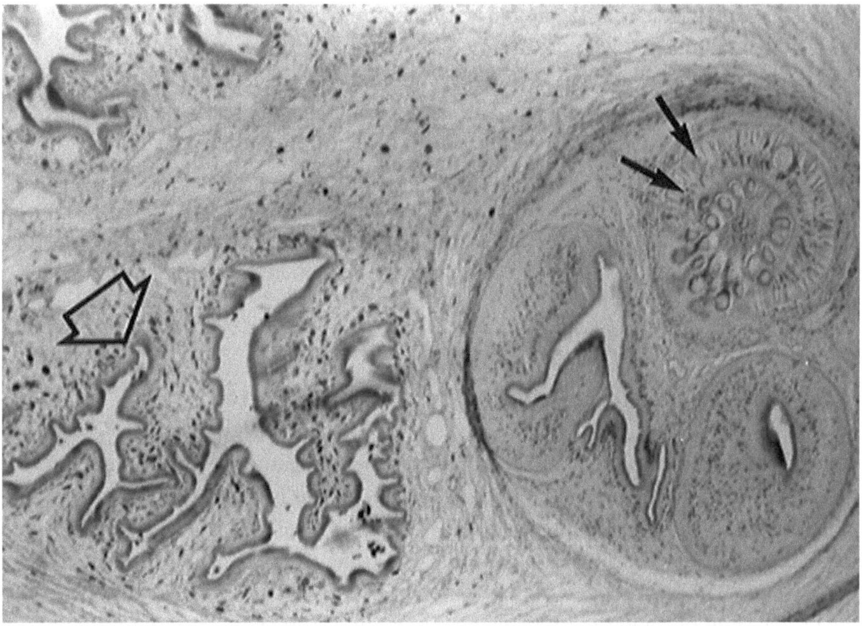

Fig. 3.2 Section of *Taenia solium* cysticercus showing spiral canal (*open arrow*) and the rostellum with characteristic hooks (*small arrows*) (From Del Brutto et al. (1998), with permission)

It is common to call *Cysticercus cellulosae* to those parasites having a scolex and *Cysticercus racemosus* to those lacking the scolex. This inadequate terminology has created confusion and it has even been considered that these are unrelated parasites originating from different *Taenia* spp. (Biagi et al. 1961). The term *Cysticercus cellulosae* was coined when it was unknown that cysticerci are the larval stage of *Taenia solium*, and the term *Cysticercus racemosus* was created to describe the appearance of some cysticerci located in the subarachnoid space that tend to group in clusters resembling a bunch of grapes. Therefore, they

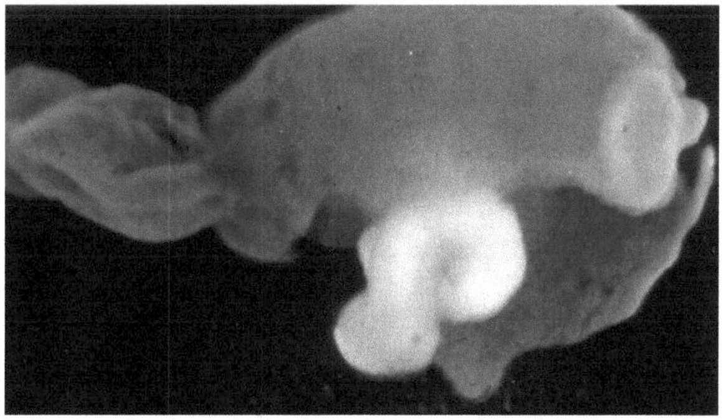

Fig. 3.3 Intermediate form of cysticercus showing typical membranes of the racemose form of cysticercus with preserved scolex (From Del Brutto et al. (1998), with permission)

should not be written as scientific names (in italics) and is preferable to describe these organisms as "the cellulose form" and "the racemose form" of cysticerci (Flisser 1994). Rabiela and coworkers (1985, 1989) described in detail the morphologic characteristics of the cellulose and the racemose forms of cysticerci, providing evidence that they belong to the same *Taenia* spp. The authors also found another form of cysticercus that conserves the scolex but has two or more small bladders sprouting from the main vesicle (Fig. 3.3). This intermediate form represents the initial stages in the transformation of the cellulose form into the racemose form of cysticercus, in which the vesicle begins to grow and the scolex disappears as the result of a degenerative process—called hydropic degeneration—that is related to the entrance of cerebrospinal fluid into the vesicles (Escobar and Weidenheim 2002).

3.2 Macroscopic Appearance of Cysticerci

The main factor determining the macroscopic appearance of cysticerci is their location in the central nervous system (Sotelo et al. 1996). Indeed, the size and shape of parasites vary according to whether they are situated in the brain parenchyma, the subarachnoid space, the ventricular system, or the spinal cord. Parenchymal brain cysticerci most often lodge in the cerebral cortex or the basal ganglia due to the high vascular supply of these areas, although some cysts may be seen in the brainstem or the cerebellum (Fig. 3.4). Most of these lesions measure from 5 to 15 mm in diameter, as the pressure of the brain parenchyma reduces the chance of further grow of the cysts. Sometimes, particularly in patients with the so-called heavy non-encephalitic form of neurocysticercosis (García and Del Brutto 1999), hundreds of cysts may be seen disseminated within the brain parenchyma, giving the brain a "Swiss-cheese" appearance (Fig. 3.5).

Subarachnoid cysticerci located at the cortical surface of the brain or within cortical sulci between two cerebral convolutions most often have a similar shape and size than parenchymal brain cysts (Fig. 3.6). According to most neuropathologists, this is the most common location of intracranial cysticerci, as many of the cysts that appear as parenchymal on neuroimaging studies are most likely

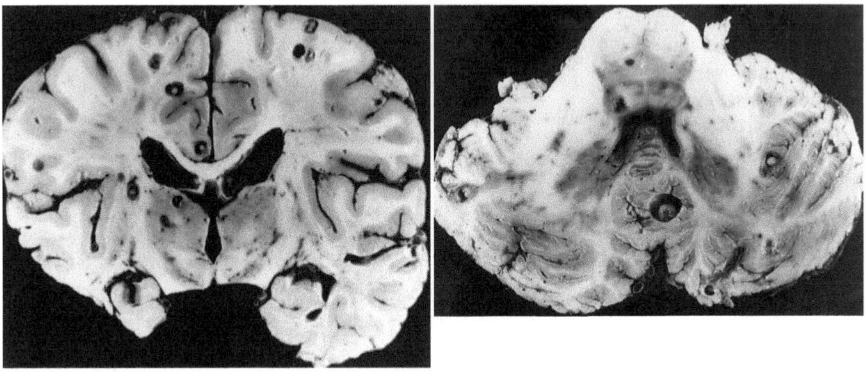

Fig. 3.4 Parenchymal brain cysticercus. Note the presence of lesions in the cerebral cortex and the basal ganglia (*left*). Some cysts are also located in the brainstem and cerebellum (*right*) (From Palmer and Reeder (2001), with permission)

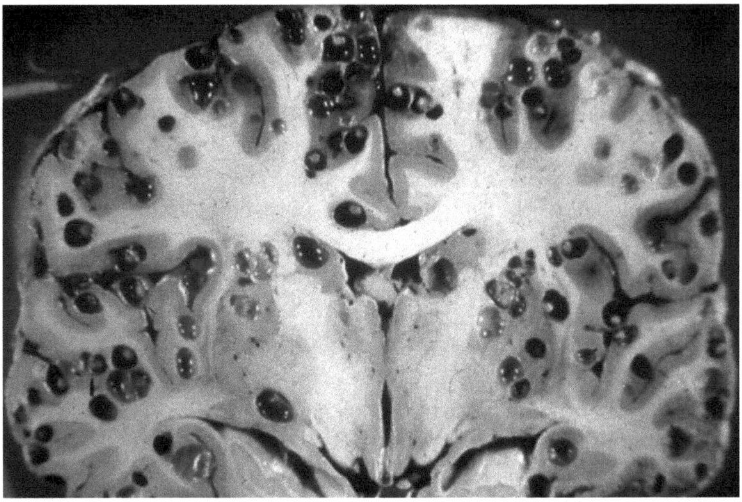

Fig. 3.5 Massive non-encephalitic parenchymal brain cysticercosis. Hundreds of living cysts (showing scolices) are predominantly located in the cerebral cortex and deep within cortical sulci (From García and Martinez (1999), with permission)

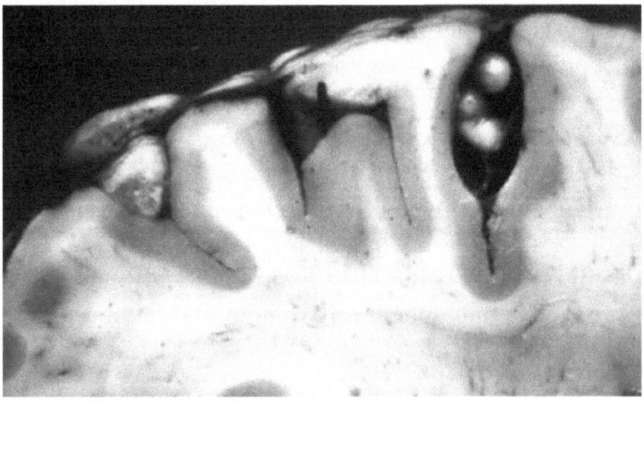

Fig. 3.6 Small subarachnoid cysticerci located in cortical sulci (From Del Brutto et al. (1998), with permission)

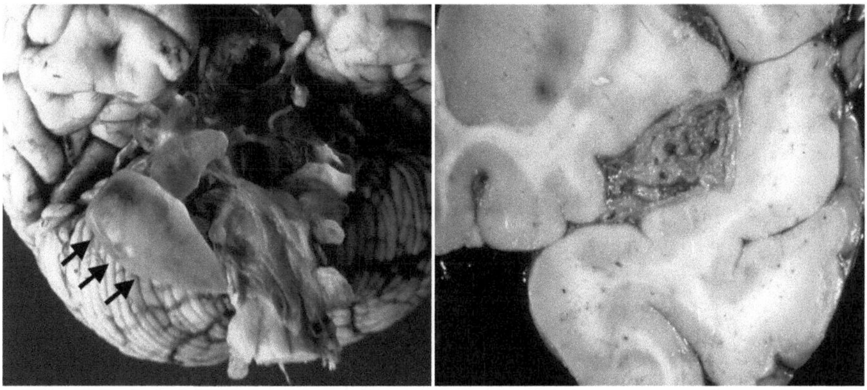

Fig. 3.7 Racemose form of neurocysticercosis located at the cerebrospinal fluid cisterns at the base of the brain (*arrows*) (*left*) and within the Sylvian fissure (*right*) (From Del Brutto et al. (1998), with permission)

subarachnoid and located deep within cortical sulci (De Souza Lino-Junior at al. 2007; Escobar et al. 1985; Kimura-Hayama et al. 2010; Pittella 1997; Rabiela-Cervantes et al. 1982). On the other hand, cysticerci located at the Sylvian fissure or within the cerebrospinal fluid cisterns at the base of the brain (including the cerebellopontine angle cistern, the ambiens and prepontine cisterns, and the optochiasmatic region) may attain a size of 50 mm or more, as their growth is not limited by the pressure effects of the brain parenchyma (Bickerstaff et al. 1952, 1956). The macroscopic appearance of these parasites is that of giant cysts associated or not with several membranes attached to each other (Fig. 3.7).

Morphologic characteristics of ventricular cysticerci are varied as they may be small or large, and may or may not have a scolex. These cysts may be found freely floating within the ventricular cavities or may be attached to the choroid plexus or the ventricular wall (Bickerstaff et al. 1956; Madrazo et al. 1983; Milenkovic et al. 1982). It is generally accepted that cysticerci enter the ventricular system through the choroid plexus of the lateral ventricles but are most often located in the fourth ventricle, though they may also be located in the third or lateral ventricles as well (Fig. 3.8). This is due to the fact that when parasites reach the ventricular cavities, they have a size that allows them to follow the cerebrospinal fluid circulation pathway from the lateral ventricles to the third ventricle through the foramina of Monro and, then, from the third to the fourth ventricle through the cerebral aqueduct. By the time cysticerci have reached the fourth ventricle, most have attained a size that prevents their further transit along the CSF circulatory pathways and get trapped within this cavity (Escobar and Nieto 1972).

Uncommon intracranial locations of cysticerci include the subdural space and the sellar region. In the few reported cases, subdural cysticerci have appeared as multilobulated cystic lesions showing scolices remnants, thus resembling

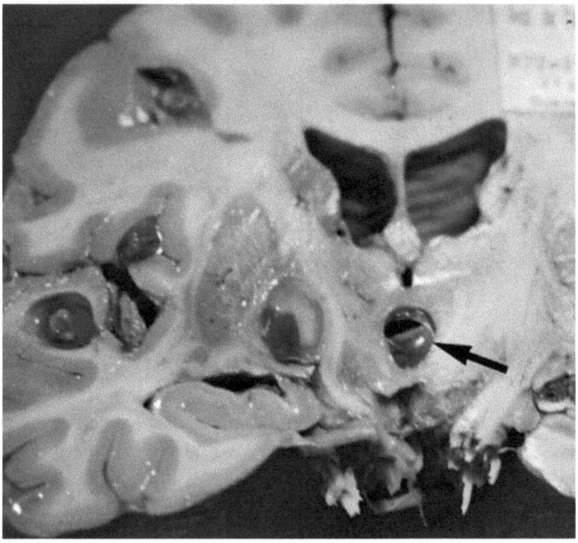

Fig. 3.8 Cysticerci located within the third ventricle (*arrow*). There are some additional parasites in the brain parenchyma (From Palmer and Reeder (2001), with permission)

those cysts located in the basal subarachnoid cisterns (Feinberg and Valdivia 1984; Im et al. 2005). Intrasellar cysticerci may occur as the result of direct sellar invasion by the parasites and may or may not be associated with subarachnoid or ventricular involvement (Arriada-Mendicoa et al. 2003; Del Brutto et al. 1988).

Cysticerci may also be found at the spinal level. In these cases, cysts may be located at the spinal subarachnoid space, the epidural space, or within the parenchyma of the spinal cord, the former accounting for about 80 % of cases of spinal cysticercosis (Gupta et al. 2009). Macroscopic appearance of spinal parasites is similar to those located within the cranial cavity (Kim and Weinberg 1985; Qi et al. 2011). Regarding the level of affection, most spinal cysticerci are seen in the thoracic region followed by the cervical, the lumbar, and the sacral regions, in this order (Guedes-Corrêa et al. 2006; Gupta et al. 2009). In some other cases, cysts may be scattered through all the length of the spinal cord (Jongwutiwes et al. 2011; Lim et al. 2010; Shin and Shin 2009). Through the years, there have been a number of hypothesis attempting to explain the occurrence of spinal cysticerci. While it has been suggested that cysticerci may enter the spinal subarachnoid and epidural spaces by retrograde flow through epidural vertebral veins (Sperlescu et al. 1989) and that intramedullary cysts may result from spread through the ventriculo-ependymal pathway (Trelles et al. 1970), it is most often accepted that spinal leptomeningeal cysticerci are the result of passive migration of cysts from the intracranial subarachnoid space, and intramedullary cysticerci are the consequence of hematogenous spread (De Souza-Queiroz et al. 1975). The former theory has been recently confirmed in a series of patients with basal subarachnoid cysticercosis whose spinal cords were evaluated with neuroimaging. In such study, 61 % of patients with basal subarachnoid cysticerci also have spinal subarachnoid space involvement (asymptomatic in many cases), suggesting that this form of the disease is more common than previously thought (Callacondo et al. 2012).

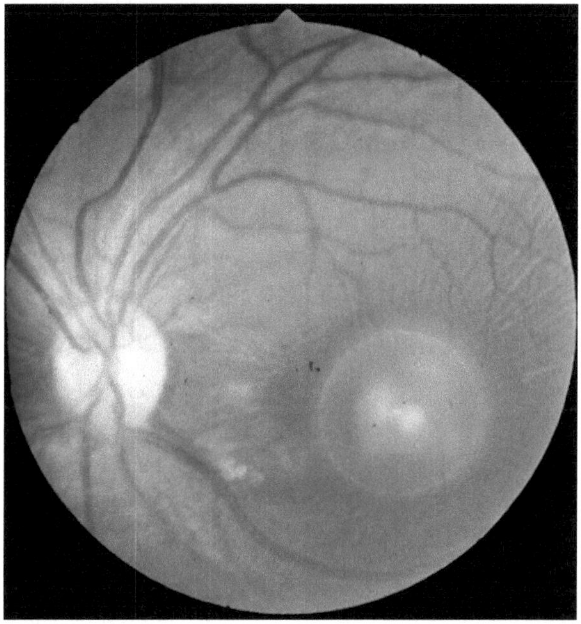

Fig. 3.9 Ophthalmoscopic appearance of subretinal cysticercus located over the macular region. The scolex is visualized as an eccentric nodule within the cyst (From Del Brutto et al. (1998), with permission)

Cysticerci may be located in the eye, including the subconjuntival space, the anterior chamber, and the subretinal space, but only the latter should be considered as neurocysticercosis since only the retina is part of the central nervous system (Malik et al. 1968; Ziaei et al. 2011). Subretinal cysts are most often located over the macular region and are supposed to enter the eye by the posterior ciliary arteries (Bartholomew 1975; Kruger-Leite et al. 1985; Lozano-Elizondo 1983). These parasites appear flattened by ophthalmoscopic examination due to the pressure effect of the vitreous against the retina. They have a white or yellowish color with a central or eccentric, well-defined, rounded zone representing the scolex (Fig. 3.9). Ocular cysticerci may also be found freely floating in the vitreous when they move from the subretinal space through the hyaloid membrane, providing the unique opportunity to visualize in vivo the evagination movements of the parasite (Lozano-Elizondo 1983; Puig-Solanes 1982). There are some rare cases of orbital cysticerci directly involving the optic nerve (Bousquet et al. 1996; Sudan et al. 2005) or extraocular muscles (Basu et al. 2009; Sudan et al. 2007).

3.3 Stages of Involution of Cerebral Cysticerci

Classical pathological studies support the concept that cysticerci are formed after the arrival of metacestodes (hexacanth embryos) to the central nervous system (Escobar 1983; Escobar and Nieto 1972). Soon after their establishment, cysticerci are viable, the bladder wall is thin and translucent, and the vesicle is filled with a clear fluid that bathes an invaginated scolex presenting a structure similar to that of

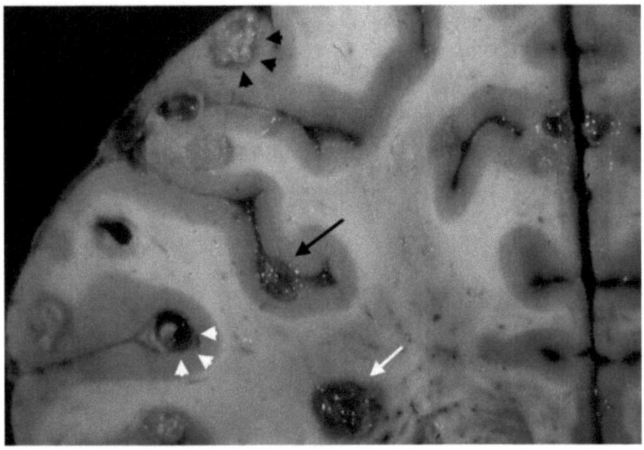

Fig. 3.10 Brain slice showing the four different stages of involution of brain cysticerci in the same person, including vesicular cyst (*white arrowheads*), colloidal cyst (*black arrow*), granuloma (*white arrow*), and calcification (*black arrowheads*) (From Sotelo et al. (1996), with permission)

the adult *Taenia solium*; this has been called the "vesicular stage." According to Escobar (1983), cysticerci may remain for years in this stage or, as the result of an immunological attack from the host, may undergo a process of degeneration that ends with the dead of the parasite and its transformation into a calcified nodule. The first stage of involution of cysticerci is the so-called colloidal stage, in which the clear vesicular fluid becomes viscous and turbid, and the scolex shows early signs of hyaline degeneration. Then, the wall of the cyst thickens and the scolex is progressively transformed into mineralized granules; this stage, in which the parasite is no longer viable, is called the "granular stage" or cysticercus granuloma (Chacko et al. 2000). Then it ensues the "calcified stage," in which cysticercus remnants appear as a solid mineralized nodule. The time parasites elapse in each of these stages has not been established, although it is accepted that considerable differences may exist among individuals (Escobar et al. 1985; Kimura-Hayama et al. 2010; Pittella 1997). Moreover, pathological studies have shown cysticerci in the four stages in the same individual, a finding that may represent cysts of different ages from recurrent infections or a single infection in which only some cysts have been attacked by the host's immune system (Fig. 3.10).

It has recently been suggested that not all cysticercus granulomas are the result of degenerated vesicular cyst that had been living for years in the brain parenchyma (going first through the vesicular and colloidal stages, respectively) but represent recently established metacestodes rapidly destroyed by the host's immune system (García et al. 2010). There are some argument favoring this alternate hypothesis of cysticerci involution, including the high rate of patients with a single cysticercus granuloma in populations exposed to low parasite loads, the low sensitivity of serological tests for the detection of anti-cysticercal antigens in patients with a single cysticercus granuloma, and the younger age of patients with a single cysticercus granuloma when compared with those with a single vesicular cyst (Del Brutto et al. 2012).

Also, recent clinical and histopathological evidence supports the fact that calcifications are not completely solid nodules and have changed previous concepts regarding the erroneous view of calcified cysticerci as totally inert lesions.

Calcifications may experience periodic morphological changes related to a mechanism of remodeling. This may expose trapped parasitic antigenic material to the host immune system, causing transient inflammatory changes in the brain parenchyma that may be the cause of recurrent seizures, focal neurological deficits or recurrent episodes of headache in some patients (Del Brutto and Del Brutto 2012; Nash et al. 2004; Ooi et al. 2011; Rathore and Radhakrishnan 2012). In addition, the increasingly accepted association and probable causal relationship between calcified cysticerci and hippocampal sclerosis favors the view that calcified lesions may cause permanent epileptogenic foci that result in the occurrence of recurrent seizures (Bianchin et al. 2010; Rathore et al. 2012).

3.4 Tissue Reactions Around Cysticerci

Changes in the nervous tissue surrounding parenchymal brain cysticerci are directly related to the stage of involution of the parasites (Table 3.1). Viable (vesicular) cysts elicit little or no inflammatory reaction in the surrounding brain parenchyma. This scarce reaction is composed of lymphocytes, plasma cells, and eosinophils and is often associated with a mild astrocytic gliosis (Thomas et al. 1989). Vesicular cysts are often surrounded by a thin collagen capsule that isolates the parasite from the host and inhibits the spontaneous evagination of the scolex arresting the evolution of the cysticercus (Ostrosky et al. 1991). On the other hand, the collagen capsule surrounding degenerating cysticerci is thick and, in turn, is surrounded by an intense inflammatory reaction that often includes the parasite itself (Fig. 3.11). A dense layer of eosinophils—playing a major role in the destruction of cysticerci—is found in contact with the external membrane of the parasite (Escobar and Weidenheim 2002; Molinari et al. 1983). The surrounding brain parenchyma also shows astrocytic gliosis, microglial proliferation, edema, neuronal degenerative changes, and perivascular cuffing of lymphocytes (Chacko et al. 2000). When parasites die, the astrocytic changes become more intense than in the

Table 3.1 Correlation between appearance of parasite and pathological changes in central nervous system according to stage of involution of parenchymal brain cysticerci

Stage of involution	Appearance of the parasite	Pathological changes in the brain parenchyma
Vesicular stage	Translucent vesicular wall; transparent vesicular fluid; viable invaginated scolex	Scarce inflammatory reaction; formation of a collagen capsule around the parasite
Colloidal stage	Thick vesicular wall; turbid vesicular fluid; scolex showing signs of hyaline degeneration	Intense inflammatory reaction that includes the parasite; thick collagen capsule around the parasite
Granular stage	Thick vesicular wall; degenerated scolex	Astrocytic gliosis around the cyst; microglial proliferation
Calcified stage	Transformation of the parasite in coarse calcified nodules	Intense gliosis; multinucleated giant cells

From Del Brutto et al. (1998), with permission

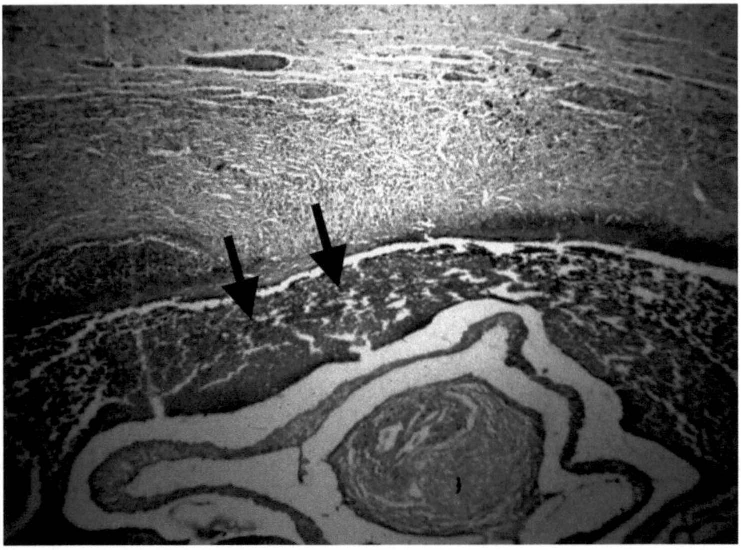

Fig. 3.11 Degenerating cysticercus surrounded by a thick collagen capsule (*arrows*). There is a severe inflammatory reaction in the surrounding brain parenchyma (From Sotelo et al. (1996), with permission)

preceding stages and is common to see gemistocytes in the surrounding tissue. As will be discussed later, this could be the pathological substrate explaining in part the malignant transformation of glial cells that is sometimes observed surrounding parenchymal brain cysticerci (Del Brutto et al. 2000). As the gliosis becomes more intense, the edema subsides, but epithelioid cells appear and coalesce to form multinucleated giant cells which are a frequent finding around calcified cysticerci (Escobar 1983; Pittella 1997).

Subarachnoid cysticerci also elicit an inflammatory reaction which, in contrast to that observed in parenchymal brain cysts, is not restricted to the perilesional tissue but is disseminated and may spread from the optochiasmatic region to the foramen magnum (Escobar and Nieto 1972). This is because activated lymphocytes, macrophages, and immunoglobulins may be carried out by the cerebrospinal fluid circulation to distant sites of the subarachnoid space. As a result, there is formation of a dense exudate composed of collagen fibers, lymphocytes, multinucleated giant cells, eosinophils, and hyalinized parasitic membranes, leading to abnormal thickening of the leptomeninges (Fig. 3.12). Such inflammatory reaction encases cranial nerves arising from the ventral aspect of the brainstem with subsequent demyelination and cranial nerve dysfunction (Fig. 3.13). The foramina of Luschka and Magendie may be occluded by the thickened leptomeninges and parasitic membranes, with the development of obstructive hydrocephalus (Estañol et al. 1983). Blood vessels arising from the circle of Willis may also be encased within this inflammatory exudate. Such changes may occur in up to 50 % of patients with subarachnoid cysticercosis but are most often asymptomatic (Barinagarrementeria

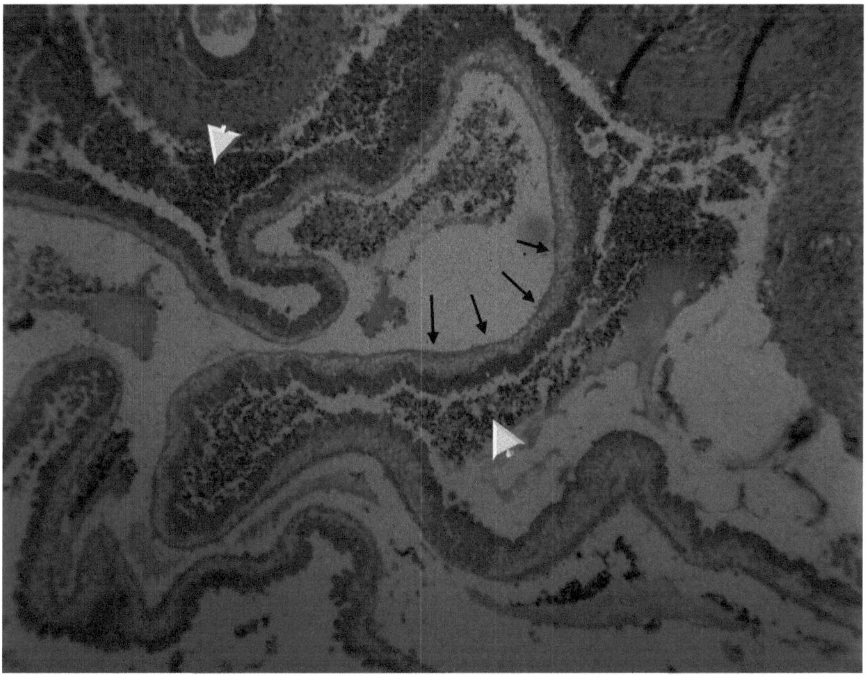

Fig. 3.12 Subarachnoid cysticercus. Parasitic membranes (*black arrows*) are embedded in a dense inflammatory exudate (*white arrowheads*)

Fig. 3.13 Racemose cysticerci in prepontine cistern surrounding cranial nerves arising from the ventral aspect of the brainstem (From Del Brutto et al. (1998), with permission)

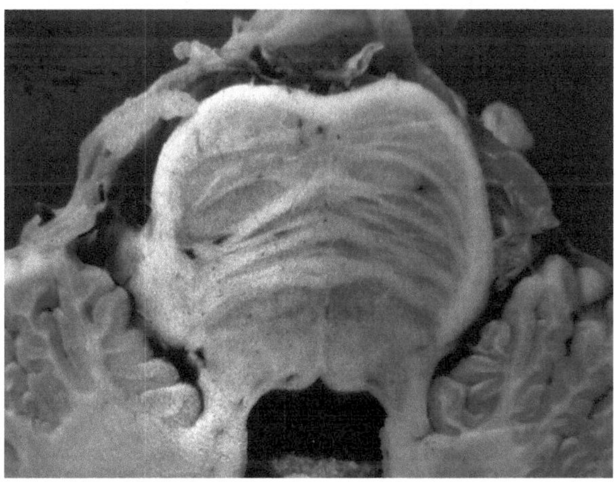

and Cantu 1998). The walls of small penetrating arteries are invaded by inflammatory cells, leading to endarteritis with thickening of the adventitia, fibrosis of the media, and endothelial hyperplasia (Fig. 3.14). The latter may reduce or occlude the lumen of the vessel with the subsequent development of a cerebral infarction (Cantú and Barinagarrementeria 1996; Del Brutto 2008; Marquez and Arauz 2012).

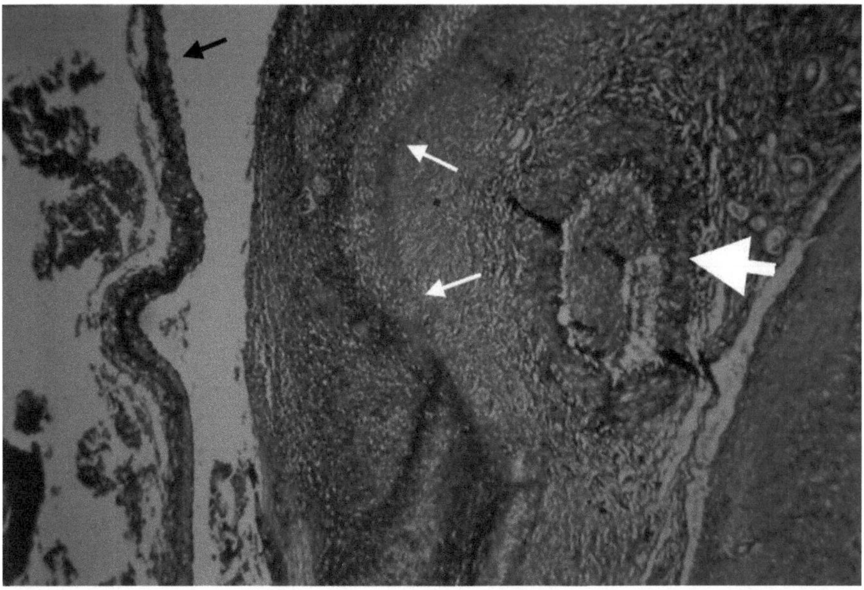

Fig. 3.14 Leptomeningeal blood vessel (*white arrowhead*) affected by endarteritis. Collagen capsule (*white arrows*) and parasitic membranes (*black arrow*) surround the vessel (Reproduced from Rodriguez-Carbajal et al. (1989), with permission)

The lumen of major intracranial arteries may be occluded by atheroma-like deposits resulting from disruption of the endothelium; this vascular involvement causes large cerebral infarcts in the territory of major intracranial arteries (Monteiro et al. 1994; Rodriguez-Carbajal et al. 1989; terPenning et al. 1992). Adherence of cysticerci to subarachnoid blood vessels may weaken the vessel wall with the subsequent development of an inflammatory (mycotic) aneurysm (Huang et al. 2000; Kim et al. 2005; Soto-Hernández et al. 1996).

Ventricular cysticerci attached to the choroid plexus or the ventricular walls are associated with an inflammatory reaction that disrupts the ependymal lining. In such cases, ependymal cells are replaced by proliferating subependymal glial cells that may protrude toward the ventricular cavities blocking the CSF transit, particularly when the site of protrusion is at or near the foramina of Monro or the cerebral aqueduct (Rangel-Guerra and Martínez 1986; Salazar et al. 1983; Siqueira et al. 1980). This process is called granular ependymitis and is associated with obstructive hydrocephalus, which may be asymmetric if only one foramina of Monro is occluded. Granular ependymitis localized in the fourth ventricle may cause syringobulbia and syringomyelia due to alterations in the hydrodynamic forces of cerebrospinal fluid circulation (Escobar and Vega 1981). In contrast, floating ventricular cysticerci that are not adhered to the ventricular walls do not elicit an inflammatory reaction; therefore, the cysts are not enclosed by the capsule that surrounds parenchymal and subarachnoid cysts. The lack of a collagen capsule makes it possible for ventricular cysts to evaginate and invaginate repeatedly, as if the parasites were searching for an appropriate place for attachment and their further transformation into adult taenias (Flisser and Madrazo 1996).

Finally, cysticerci located in the spinal subarachnoid space may induce inflammatory or demyelinating changes in ventral and dorsal roots, in a similar way as intracranial subarachnoid cysts induce damage in cranial nerves. Cysticerci located inside the spinal cord parenchyma induce chronic inflammatory changes resembling those described in the brain parenchyma (Agale et al. 2012; Mohanty et al. 1997; Qi et al. 2011).

References

Agale SV, Bhavsar S, Choudhury B, Manohar V (2012) Isolated intramedullary spinal cord cysticercosis. Asian J Neurosurg 7:90–92

Arriada-Mendicoa N, Celis-López MA, Higuera-Calleja J, Corona- Vázquez T (2003) Imaging features of sellar cysticercosis. AJNR Am J Neuroradiol 24:1386–1389

Barinagarrementeria F, Cantú C (1998) Frequency of cerebral arteritis in subarachnoid cysticercosis: an angiographic study. Stroke 29:123–125

Bartholomew RS (1975) Subretinal cysticercosis. Am J Ophthalmol 79:670–673

Basu S, Muthusami S, Kumar A (2009) Ocular cysticercosis: an unusual cause of ptosis. Singapore Med J 50:e309–e311

Biagi FF, Briceño CE, Martínez B (1961) Diferencias entre *Cysticercus cellulosae* y *Cysticercus racemosus*. Rev Biol Trop (Méx) 9:141–151

Bianchin MM, Velasco TR, Wichert-Ana L, Takayanagui OM, Leite JP, Sakamoto AC (2010) How frequent is the association of neurocysticercosis and mesial temporal lobe epilepsy with hippocampal sclerosis. Epilepsia 51:2359–2360

Bickerstaff ER, Cloake PCP, Hughes B, Smith WT (1952) The racemose form of cerebral cysticercosis. Brain 75:1–18

Bickerstaff ER, Small JM, Woolf AL (1956) Cysticercosis of the posterior fossa. Brain 79: 622–634

Bousquet CF, Dufour TFL, Derome PCE (1996) Retrobulbar optic nerve cysticercosis. Case report. J Neurosurg 84:293–296

Callacondo D, García HH, Gonzales I, Escalante D, Nash TE (2012) High frequency of spinal involvement in patients with basal subarachnoid neurocysticercosis. Neurology 78:1394–1400

Cantú C, Barinagarrementeria F (1996) Cerebrovascular complications of neurocysticercosis. Clinical and neuroimaging spectrum. Arch Neurol 53:233–239

Chacko G, Rajshekhar V, Chandy MJ, Chandi SM (2000) The calcified intracorporeal vacuole: an aid to the pathological diagnosis of solitary cerebral cysticercus granuloma. J Neurol Neurosurg Psychiatry 69:525–527

De Souza Lino-Junior R, Guimarães Faleiros AC, Vinaud MC, de Oliveira FA, Valadares Guimãraes J, dos Reis MA, de Paula Antunes Teixeira V (2007) Anatomopathological aspects of neurocysticercosis in autopsied patients. Arq Neuropsiquiatr 65:87–91

De Souza-Queiroz L, Filho AP, Callegaro D, De Faria LL (1975) Intramedullary cysticercosis: case report, literature review, and comments on pathogenesis. J Neurol Sci 26:61–70

Del Brutto OH (2008) Stroke and vasculitis in patients with neurocysticercosis. In: Caplan LR (ed) Uncommon causes of stroke. Cambridge University Press, Cambridge/New York

Del Brutto OH, Del Brutto VJ (2012) Calcified neurocysticercosis among patients with primary headache. Cephalalgia 32:250–254

Del Brutto OH, Guevara J, Sotelo J (1988) Intrasellar cysticercosis. J Neurosurg 69:58–60

Del Brutto OH, Sotelo J, Del Brutto OH (eds) (1998) Neurocysticercosis. A clinical handbook. Swets & Zeitlinger Publishers, Lisse

Del Brutto OH, Dolezal M, Castillo PR, García HH (2000) Neurocysticercosis and oncogénesis. Arch Med Res 31:151–155

Del Brutto VJ, Del Brutto OH, Ochoa E, Garcia HH (2012) Single parenchymal brain cysticercus: relationship between age of patients and evolutive stage of parasites. Neurol Res 34:967–970

Escobar A (1983) The pathology of neurocysticercosis. In: Palacios E, Rodriguez-Carbajal J, Taveras JM (eds) Cysticercosis of the central nervous system. Charles C. Thomas Publisher, Springfield

Escobar A, Nieto D (1972) Parasitic diseases. In: Minckler J (ed) Pathology of the nervous system, vol 3. McGraw-Hill Book Company, New York

Escobar A, Vega J (1981) Syringomyelia and syringobulbia secondary to arachnoiditis and fourth ventricle blockage due to cysticercosis. A case report. Acta Neuropathol (Berl) Suppl VII:389–391

Escobar A, Weidenheim KM (2002) The pathology of neurocysticercosis. In: Singh G, Prabhakhar S (eds) *Taenia solium* cysticercosis. From basic to clinical science. CABI Publishing, Oxon

Escobar A, Aruffo C, Cruz-Sánchez F, Cervos-Navarro J (1985) Hallazgos neuropatológicos en la neurocisticercosis. Arch Neurobiol (Madrid) 48:151–156

Estañol B, Kleriga E, Loyo M, Mateos H, Lombardo L, Gordon F, Saguchi AF (1983) Mechanisms of hydrocephalus in cerebral cysticercosis: implications for therapy. Neurosurgery 13:119–123

Feinberg WM, Valdivia FR (1984) Cysticercosis presenting as a subdural hematoma. Neurology 34:1112–1113

Fleury A, Escobar A, Fragoso G, Sciutto E, Larralde C (2010) Clinical heterogeneity of human neurocysticercosis results from complex interactions among parasite, host and environmental factors. Trans R Soc Trop Med Hyg 104:243–250

Flisser A (1994) Taeniasis and cysticercosis due to *Taenia solium*. In: Sun T (ed) Progress in clinical parasitology. CRC Press, Boca Raton

Flisser A, Madrazo I (1996) Evagination of *Taenia solium* in the fourth ventricle. N Engl J Med 335:753–754

García HH, Del Brutto OH (1999) Heavy nonencephalitic cerebral cysticercosis in tapeworm carriers. Neurology 53:1582–1584

García HH, Martinez SM (eds) (1999) *Taenia solium* taeniasis/cysticercosis. Editorial Universo, Lima

García HH, Gonzalez AE, Rodriguez S, Tsang VC, Pretell EJ, Gonzales I, Gilman RH (2010) Neurocysticercosis. Unraveling the nature of the single cysticercal granuloma. Neurology 75:654–658

Guedes-Corrêa JF, Caratta Macedo R, Pereira Vaitsman R, Gomes de Mattos J, Marques Agra J (2006) Intramedullary spinal cysticercosis simulating a conus medullary tumor. Arq Neuropsiquiatr 64:149–152

Gupta S, Singh PK, Gupta B, Singh V, Azam A (2009) Isolated primary intradural extramedullary spinal cysticercosis: a case report and review of the literature. Acta Neurol Taiwan 18:187–192

Huang PP, Choudhri HF, Jallo G, Douglas M (2000) Inflammatory aneurysm and neurocysticercosis: further evidence for a causal relationship? Case report. Neurosurgery 47:466–467

Im S-H, Park S-H, Oh DH, Kang B-S, Kwon O-K, Oh CW (2005) Subdural cysticercosis mimicking a chronic subdural hematoma. J Neurosurg 102:389

Jongwutiwes U, Yanagida T, Ito A, Kline SE (2011) Isolated intradural-extramedullary spinal cysticercosis: a case report. J Travel Med 18:284–287

Kim KS, Weinberg PE (1985) Spinal cysticercosis. Surg Neurol 24:80–82

Kim I-Y, Kim T-S, Lee J-H, Lee M-C, Lee J-K, Jung S (2005) Inflammatory aneurysm due to neurocysticercosis. J Clin Neurosci 12:585–588

Kimura-Hayama ET, Higuera JA, Corona-Cedillo R, Chávez-Macías L, Perochena A, Quiroz-Rojas LY, Rodriguez-Carbajal J, Criales JL (2010) Neurocysticercosis: radiologic-pathologic correlation. Radiographics 30:1705–1719

Kruger-Leite E, Jalkh AE, Quiroz H, Schepens CL (1985) Intraocular cysticercosis. Am J Ophthalmol 99:252–257

Lim BC, Lee RS, Lim JS, Cho KY (2010) A case of neurocysticercosis in entire spinal level. J Korean Neurosurg 48:371–374

Lozano-Elizondo D (1983) Ophthalmic cysticercosis. In: Palacios E, Rodriguez-Carbajal J, Taveras JM (eds) Cysticercosis of the central nervous system. Charles C. Thomas Publisher, Springfield

Madrazo I, García-Rentería JA, Sandoval M, López-Vega FJ (1983) Intraventricular cysticercosis. Neurosurgery 12:148–152

Mahanty S, García HH (2010) Cysticercosis and neurocysticercosis as pathogens affecting the nervous system. Prog Neurobiol 91:172–184

Malik SRK, Gupta AK, Choudhry S (1968) Ocular cysticercosis. Am J Ophthalmol 66: 1168–1171

Marquez JM, Arauz A (2012) Cerebrovascular complications of neurocysticercosis. Neurologist 18:17–22

Milenkovic Z, Penev G, Stojanovic D, Jovicic V, Antovic P (1982) Cysticercosis cerebri involving the lateral ventricle. Surg Neurol 18:94–96

Mohanty A, Venkatrama SK, Das S, Das BS, Rao BR, Vasudev MK (1997) Spinal intramedullary cysticercosis. Neurosurgery 40:82–87

Molinari JL, Meza R, Tato P (1983) *Taenia solium*: cell reactions to the larva (*Cysticercus cellulosae*) in naturally parasitized immunized hogs. Exp Parasitol 56:327–338

Monteiro L, Almeida-Pinto J, Leite I, Xavier J, Correia M (1994) Cerebral cysticercus arteritis: five angiographic cases. Cerebrovasc Dis 4:125–133

Nash TE, Del Brutto OH, Butman JA, Corona T, Delgado-Escueta A, Duron RM, Evans CA, Gilman RH, Gonzalez AE, Loeb JA, Medina MT, Pietsch-Escueta S, Pretell EJ, Takayanagui OM, Theodore W, Tsang VC, García HH (2004) Calcific neurocysticercosis and epileptogenesis. Neurology 62:1934–1938

Ooi WW, Wijemanne S, Thomas CB, Quezado M, Brown CR, Nash TE (2011) A calcified *Taenia solium* granuloma associated with recurrent perilesional edema causing refractory seizures: histopathological features. Am J Trop Med Hyg 85:460–463

Ostrosky L, Correa D, Faradi R, García H, Flisser A (1991) *Taenia solium* inhibition of spontaneous evagination of cysticerci by the host inflammatory capsule. Int J Parasitol 21:603–604

Palmer PES, Reeder MM (2001) The imaging of tropical diseases, 2nd edn. Springer, Heidelberg

Pittella JEH (1997) Neurocysticercosis. Brain Pathol 7:681–693

Puig-Solanes M (1982) Algunos aspectos anátomo-clínicos del cisticerco intraocular. Salud Púb Méx 24:649–660

Qi B, Ge P, Yang H, Bi C, Li Y (2011) Spinal intramedullary cysticercosis: a case report and literature review. Int J Med Sci 8:420–423

Rabiela MT, Rivas A, Castillo S, González-Angulo A (1985) Pruebas morfológicas de que *C. cellulosae* y *C. racemosus* son larvas de *T. solium*. Arch Invest Med Méx 16:83–86

Rabiela MT, Rivas A, Flisser A (1989) Morphological types of *Taenia solium* cysticerci. Parasitol Today 5:357–359

Rabiela-Cervantes MT, Rivas-Hernandez A, Rodriguez-Ibarra J, Castillo-Medina S, Cancino FM (1982) Anatomopathological aspects of human brain cysticercosis. In: Flisser A, Willms K, Laclete JP, Larralde C, Ridaura C, Beltrán F (eds) Cysticercosis: present state of knowledge and perspectives. Academic, New York

Rangel-Guerra R, Martínez HR (1986) Diagnóstico diferencial de la estenosis del acueducto de Silvio en el adulto. Rev Invest Clin (Méx) 38:21–27

Rathore C, Radhakrishnan K (2012) What causes seizures in patients with calcified neurocysticercal lesions. Neurology 78:612–613

Rathore C, Thomas B, Kesavadas C, Radhakrishnan K (2012) Calcified neurocysticercosis and hippocampal sclerosis: potential dual pathology? Epilepsia 53:e60–e62

Rodriguez-Carbajal J, Del Brutto OH, Penagos P, Huebe J, Escobar A (1989) Occlusion of the middle cerebral artery due to cysticercotic angiitis. Stroke 20:1095–1099

Salazar A, Sotelo J, Martinez H, Escobedo F (1983) Differential diagnosis between ventriculitis and fourth ventricle cyst in neurocysticercosis. J Neurosurg 59:660–663

Shin DA, Shin HC (2009) A case of extensive spinal cysticercosis involving the whole spinal canal in a patient with a history of cerebral cysticercosis. Yonsei Med J 50:582–584

Siqueira EB, Richardson RR, Kranzler LI (1980) Cysticercosis occluding the foramen of Monro. Surg Neurol 13:429–431

Sotelo J, Del Brutto OH, Roman GC (1996) Cysticercosis. In: Remington JS, Swartz MN (eds) Current clinical topics in infectious diseases, vol 16. Blackwell Science, Inc., Boston

Soto-Hernández JL, Gomez-Llata S, Rojas-Echeverri LA, Texeira F, Romero V (1996) Subarachnoid hemorrhage secondary to a ruptured inflammatory aneurysm: a possible manifestation of neurocysticercosis: case report. Neurosurgery 38:197–200

Sperlescu A, Balbo RJ, Rossitti SL (1989) Breve comentário sobre a patogenia da cisticercose espinhal. Arq Neuropsiquiatr 47:105–109

Sudan R, Muralidhar R, Sharma P (2005) Optic nerve cysticercosis: case report and review of current management. Orbit 24:159–162

Sudan AR, Mouzinho A, Valente P (2007) Orbital cysticercosis: diagnosis and treatment controversies. Pediatr Infect Dis J 26:180–181

Tay-Zavala J (1983) Etiology of cysticercosis. In: Palacios E, Rodriguez-Carbajal J, Taveras JM (eds) Cysticercosis of the central nervous system. Charles C. Thomas Publisher, Springfield

terPenning B, Litchmanm CD, Heier L (1992) Bilateral middle cerebral artery occlusions in neurocysticercosis. Stroke 23:280–283

Thomas JA, Knoth R, Schwechheimer K, Volk B (1989) Disseminated human neurocysticercosis. Acta Neuropathol 78:594–604

Trelles JO, Caceres A, Palomino L (1970) La cysticercose medullaire. Rev Neurol (Paris) 123:187–202

Valkounova J, Zdarska Z, Slais J (1992) Histochemistry of the racemose form of *Cysticercus cellulosae*. Folia Parasitol Praha 39:207–226

Ziaei M, Elgohary M, Bremmer FD (2011) Orbital cysticercosis, case report and review. Orbit 30:230–235

Epidemiology of Human Cysticercosis in Endemic Regions

4

Taenia solium establishes its life cycle in regions where domestic pig raising coexists with poor sanitary conditions. Unfortunately, most poor areas of the world, with exception of Muslim populations, fulfill these simple requirements (Gilman et al. 1999). In poor regions, domestic pig raising is a usual component of survival economies, driven as a major incentive by the fact that pigs can be just left to roam by themselves and thus no money or effort need to be spent in food or forage, resulting in a very cheap production. Furthermore, there are very active commercialization systems for pigs and pork in most towns (Cysticercosis Working Group in Peru 1993).

The World Health Organization has drawn an official map showing areas where *Taenia solium* cysticercosis is endemic (Fig. 4.1). These include most Latin

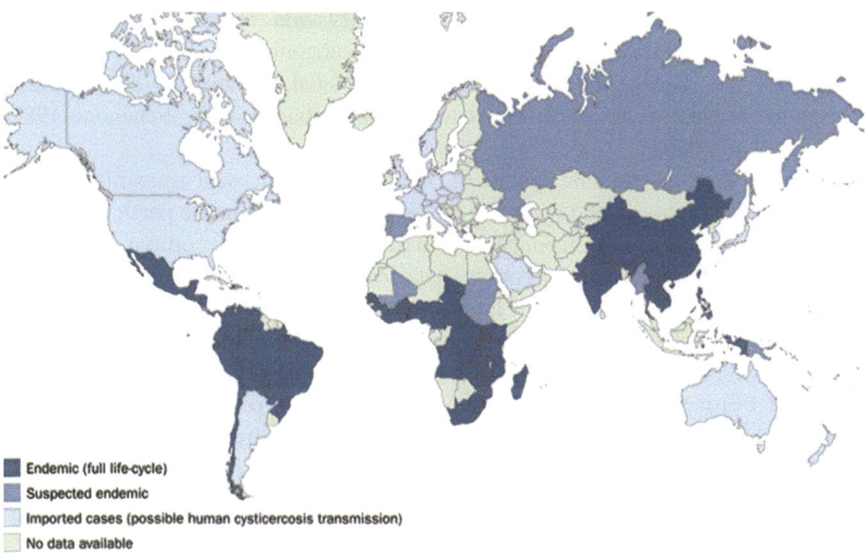

Fig. 4.1 World map showing areas where cysticercosis is endemic

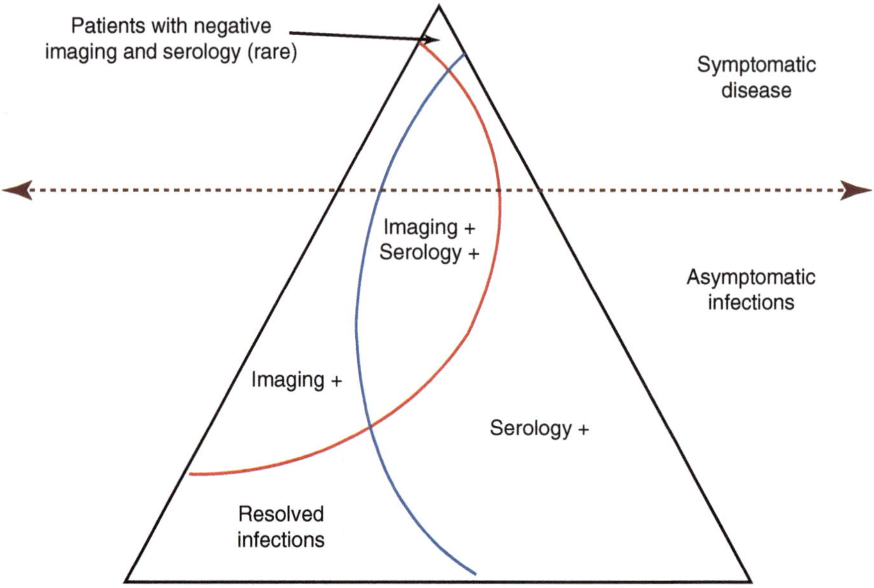

Fig. 4.2 Diagram showing gross prevalence of symptomatic and asymptomatic cysticercosis infections according to positive or negative results on neuroimaging studies and serology

American countries with the exception of Argentina and Uruguay, some islands of the Caribbean basin including Haiti and Dominican Republic, the sub-Saharan Africa (Benin, Burkina Faso, Burundi, Cameroon, Chad, Congo, Ghana, Kenya, Madagascar, Mozambique, Nigeria, Rhodesia, Rwanda, South Africa, Tanzania, Togo, Uganda, Zambia, Zimbabwe), the Indian subcontinent, and vast regions of Southeast Asia (Burma, China Laos, Indonesia, Nepal, Philippines, Thailand, and Vietnam) (O'Neal et al. 2012; Preux and Druet-Cabanac 2005; Schenone 1982; Winkler et al. 2010).

Even in disease-endemic countries, marked differences in transmission and prevalence of Taeniasis and cysticercosis exist depending on the particular population group studied. Data from rural areas is not necessarily comparable to data from urban metropolis, and of course, data from hospital-based registries should not be compared with data from the general population (García and Del Brutto 2005). In a given population, there will be a series of diverse subgroups of infected persons in terms of symptomatic cases or serological or imaging evidence of cysticercosis (Fig. 4.2). On this basis, it is possible to differentiate among infected and noninfected individuals and on whether infections are symptomatic or not (Table 4.1). There are some limitations in this categorization, mainly arising from the fact that imaging is mostly restricted to diagnose neurocysticercosis.

Table 4.1 Subgroups of infected persons in terms of symptomatic cases or serological or neuroimaging evidence of cysticercosis

	Not infected	Infected	
		Asymptomatic	Symptomatic
Negative for antibody, antigen, and imaging	+		
Exposed: antibody positive, antigen negative, CT negative		+	+
Infected, no viable lesions: calcifications on neuroimaging, antigen negative (independent of antibody results)		+	+
Infected, degenerating lesions: transitional, enhancing cysts on neuroimaging (often a single lesion)		±	+
Infected, viable lesions: cysts on neuroimaging or positive antigen, often antibody positive		+	+

4.1 Data on Asymptomatic Individuals/General Population

4.1.1 Rates of Infection

As noted, assessing the prevalence of cysticercosis may be difficult due to the high percentage of asymptomatic infections. The most common ways to evaluate prevalence in rural endemic populations are by serology (anticysticercal antibodies or cysticercal antigens detection) or, rarely, by neuroimaging.

4.1.1.1 Anticysticercal Antibody Prevalence

Many surveys exist in the literature and of course the estimates vary according to the sensitivity and specificity of the test used. Before the introduction of the enzyme-linked immunoelectrotransfer blot (EITB), the low sensitivity and specificity of the existing tests made it difficult to assess the extent of infection in endemic regions. Woodhouse et al. (1982) reported an overall 1 % of seropositivity in 20,000 samples from Mexico; with some communities reaching 6 %, the sensitivity and specificity of the test used in that survey—immunoelectrophoresis—were estimated to be around 50 %. Thereafter, some population-based studies used the enzyme-linked immunosorbent assay (ELISA) with varied results. Coker-Vann et al. (1981) reported seroprevalence rates ranging from 0.2 to 8.4 % in asymptomatic individuals living in different regions of Indonesia, which contrasted with the 16 % seroprevalence rate of asymptomatic individuals living around the Wissel Lakes area, where cysticercosis was highly endemic. In the latter, patients with diagnosis of neurocysticercosis had a seroprevalence of 61 %. In another study performed in rural Madagascar, 18 % of healthy individuals versus 36 % of neurocysticercosis patients had a positive ELISA in serum (Michel et al. 1993). Posterior studies, showing the disappointing reliability of the serum ELISA, make these surveys of historical importance only.

Once the highly specific EITB using purified lentil lectin glycoprotein antigens was available, multiple surveys have demonstrated that from 8 to 25 % of "healthy" persons living in endemic villages have antibodies to *Taenia solium* (García et al. 1991; Sarti et al. 1992). Asymptomatic seropositive individuals mostly present weak reactions to only one to three antibody bands (GP50, GP42-39, and GP24) (Schantz et al. 1994). In field conditions, antibody seroprevalence rates may overestimate the actual prevalence of infection because persons with anticysticercal antibodies resulting from exposure or from past infections are also detected. Nevertheless, antibody prevalence can give a general assessment of the levels of transmission and help to orientate control strategies. At the individual level, detection of specific anticysticercal antibodies in an asymptomatic person has limited clinical use.

4.1.1.2 Cysticercal Antigens Prevalence

More recently, some population-based data using circulating cysticercal antigen detection has become available. The prevalence of circulating antigen ranges from 1 to 22 % (Dorny et al. 2012; Mwanjali et al. 2013; Mwape et al. 2013; Nguekam et al. 2003; Praet et al. 2010). Antibody prevalence is usually much higher than antigen prevalence in these studies. There is no information about the predictive value of a positive antigen test in asymptomatic individuals or in the general population, neither is it known whether the use of a higher cutoff would improve its predictive value for neurocysticercosis. As for community-based antibody surveys, the application of antigen detection in an asymptomatic individual would likely be of no practical use, although its use for early detection of extraparenchymal infections has been suggested (García et al. 2012).

4.1.1.3 Prevalence Using Neuroimaging

The proportion of asymptomatic individuals with neurocysticercosis detected by computed tomography (CT) in endemic populations ranges from 10 to 25 % (Del Brutto et al. 2005; Fleury et al. 2003; Garcia-Noval et al. 1996, 2001; Medina et al. 2005; Montano et al. 2005; Sánchez et al. 1999) (Table 4.2). A single study in a high-risk population using magnetic resonance imaging (MRI) reported 5 % of people harboring small viable cysts, usually a single one (Prasad et al. 2011)

A closer view to the results showed in Table 4.2 provides useful insights. Field studies using CT consistently demonstrate a significant proportion of villagers harboring one or a few brain calcifications, more rarely multiple calcifications or even more rarely viable cysts. In these same field conditions, most asymptomatic, antibody-positive people show weak reactions. The positive predictive value of a positive antibody test in asymptomatic individuals is around 35 %, and the imaging correlate in such cases is usually one or a few parenchymal brain calcifications. The relative frequency of viable cysts in general population is low. Not all cases of neurocysticercosis are detected by an antibody assay, since a proportion of seronegative villagers will also show calcifications on brain CT (Montano et al. 2005). Overall, approximately 50 % of asymptomatic individuals with parenchymal brain

Table 4.2 Epidemiologic surveys evaluating the proportion of asymptomatic infections/symptomatic neurocysticercosis (documented by neuroimaging) among persons with epilepsy in endemic areas

Location (author)	Proportion with NCC on neuroimaging		Ratio of NCC in individuals with epilepsy/general population
	Asymptomatic individuals	People with seizures or with epilepsy	
Guatemala (Garcia-Noval et al. 1996)	12/51 (24 %)	36/76 (47 %)	2.01
Ecuador (Cruz et al. 1999)	17/118 (14 %)	14/26 (54 %)	3.74
Honduras (Sánchez et al. 1999)	29/144 (20 %)	2/4 (50 %)	2.50
Mexico (Fleury et al. 2003)	14/153 (9 %)	0/1	…
Peru (Montano et al. 2005)	Seronegative: 8/58 (14 %) Seropositive: 18/53 (34 %)	15/39 (38 %)	2.00[a]
Ecuador (Del Brutto et al. 2005)	1/19 (5 %)	5/19 (26 %)	5.00
Tanzania (Winkler et al. 2009)	10/198 (5 %)	38/212 (18 %)	3.54

[a]Calculated over a weighted average at 24.2 % seroprevalence in general population

calcifications are EITB negative. A proportion of villagers will have a positive antigen test, apparently fewer than those who are antibody positive. The positive predictive value of a positive antigen test in an asymptomatic individual is not well known and has been reported to be from 14 to 53 % (Mwanjali et al. 2013; Nguekam et al. 2003). The neuroimaging correlation for a positive result in antigen detection is usually that of multiple viable parenchymal brain cysts. It is unclear whether the remaining antigen-positive, neuroimaging-negative individuals may have viable brain cysts not detected by imaging or cysts elsewhere in their body.

4.1.1.4 Factors Associated to Infection in Field Settings

Diverse epidemiological studies and clinical series have demonstrated associations between positive tests and a history of intestinal taeniasis, pig raising, having seen cysticercosis in their pigs, and other more generic factors such as living in a rural area and poor sanitation. A history of intestinal taeniasis in the person or in the household is a very strong risk factor. The accuracy of questioning for an antecedent of taeniasis could be improved by showing villagers a proglottid, since patients who had expelled nematodes may assume that the question refers to this antecedent and answer affirmatively. Pig raising is a poorly specific indicator since in many villages more than half of the villagers raise pigs (Diaz et al. 1992; Garcia-Noval et al. 1996; Sarti et al. 1994). Having seen cysticercosis cysts in pigs (or in their own pigs) could provide a more specific marker, but we have not been able to find comparative assessments of the predictive value of these questions. There seems not to be particularly clear differences in the rates of infection be gender. In contrast, the prevalence of antibody-positive individuals in the general population begins to rise between 6 and 10 years of age and increase progressively with higher rates between 20 and 50 years of age (Diaz et al. 1992; Schantz et al. 1994).

4.1.1.5 Geographic Pattern/Clustering

Clustering of human and porcine cases around a tapeworm carrier was first described by Sarti-Gutierrez et al. (1988). Using GPS, Lescano et al. (2007) confirmed this and described that a stronger concentrically gradient is found for porcine seroincidence as an indicator of more recent transmission. Seizure cases, however, are not clustered around the tapeworm carrier, likely reflecting a clinically silent period between infection and seizures longer than the average life span of the adult tapeworm (Lescano et al. 2009). Familiar aggregation of cases is likely due to common source exposure, although some sort of genetic predisposition to infection or disease cannot be ruled out. The polymorphism of toll-like receptor 4 has been shown to be associated to symptomatic disease in human neurocysticercosis (Verma et al. 2010), and differences in some HLA determinants between patients and controls were demonstrated long ago in Mexico (Del Brutto et al. 1991).

4.1.1.6 Dynamics of Infection in Endemic Populations

To better interpret antigen or antibody prevalence surveys, it is important to note that a significant proportion of reactors will switch from seropositive to seronegative in 1 year or less (García et al. 2001; Meza-Lucas et al. 2003; Mwape et al. 2013). These individuals correspond to weaker immune reactions and likely represent exposure without infection, reexposure, mild infections, or infections which resolved spontaneously. It follows that in endemic communities exposure and transmission are dynamic processes which likely involve a large proportion of the population along the years.

4.1.1.7 Cysticercosis in Necropsy Series

Cysticercosis has been reported in 1–3 % of necropsies in general hospitals of Mexico, Peru, Brazil, and other countries. These cases were mostly cysts in brain, heart, or other systematically examined organs (Briceño et al. 1961; Escalante 1977; Schenone 1982; Lino-Junior et al. 2007). These findings were reported as reflecting that 50 % or more of all cases were asymptomatic. However, while necropsy examination demonstrated that cysticercosis was not the cause of death, it confers no certainty that these individuals were truly asymptomatic.

4.2 Data on Symptomatic Patients/Clinical Series

Information on the prevalence of cysticercosis based on hospital-based studies is even less reliable than that obtained from field surveys. Moreover, the proportion of the different forms of neurocysticercosis is unrealistic, since more severe cases are those to seek medical attention. Nevertheless, some valuable information can be gathered from such studies.

4.2.1 Antibodies, Antigen, and Imaging in Clinical Series

Published clinical series are generally reported from neurological services (most often series of patients with seizures or epilepsy) or neurosurgical services. These clinical series show most patients with moderate to strong antibody reactions (except for a proportion of individuals with only calcified disease or a single brain cyst), positive antigen reactions (less sensitive in individuals with only one or a few parenchymal cysts, negative in individuals with only calcifications), and neuroimaging findings involving viable cysts and subarachnoid disease in most patients, except in the Indian subcontinent where the vast majority of neurocysticercosis patients have a single enhancing brain lesion (García et al. 1997; Rajshekhar and Chandy 2000; Wilson et al. 1991). Almost all series refer seizures as the more frequent clinical manifestation, ranging from 70 to 90 % of cases (Del Brutto et al. 1992). Extraparenchymal neurocysticercosis, however, is found with low frequency in neurological services. Its proportional contribution varies from 10 % in neurology services to over 60 % in neurosurgical series, reflecting the impact in hydrocephalus and mass lesions which require surgical management (Fleury et al. 2011).

4.2.2 Rates of Disease

How much symptomatic disease can be attributed to neurocysticercosis has been a matter of unnecessary debate. Its frequency in clinical series exceeds by far what could be expected in the general population. This was initially demonstrated in controlled serological studies by Chopra et al. (1981) in India using hemagglutination. These findings were consistently replicated as new, improved serological assays were available (García et al. 1991; Rosas et al. 1986). The proportions of neuroimaging findings compatible with neurocysticercosis in individuals with epilepsy in endemic regions may reach 50 % (Cruz et al. 1999; García et al. 1999; Medina et al. 1990; Sánchez et al. 1999). Once the serological and imaging data are consolidated, consistent estimates of approximately 30 % of individuals with epilepsy in endemic regions are attributable to neurocysticercosis (Del Brutto et al. 2005; Montano et al. 2005; Ndimubanzi et al. 2010)

There is no systematic assessment of what proportion of individuals with obstructive hydrocephalus or intracranial hypertension are due to neurocysticercosis, although its contribution to hydrocephalus series seems quite high (Lobato et al. 1981; Suástegui-Roman et al. 1996). On the other hand, vascular abnormalities seem to be frequent in cases of neurocysticercosis, and it clearly can be a cause of stroke; however, the proportional contribution of neurocysticercosis to the burden of symptomatic strokes in endemic countries does not seem to be major (Del Brutto 1992; Barinagarrementeria and Cantu 2002).

4.2.3 Factors Associated to Infection and Disease

In clinical settings, raising pigs and a history of taeniasis have some predictive value (García et al. 1995). From published series, from 5 to 10 % of all neurocysticercosis patients carry an intestinal tapeworm by the time of neurological symptoms, and around 25 % refer a history of having passed tapeworm segments before (Gilman et al. 2000). Intestinal taeniasis is almost always present in individuals with massive infections. A tapeworm carrier in the household is a very common finding when cysticercosis presents in infants and young children (Del Brutto 2012), probably reflecting the short possible time elapsed from exposure. Disease characteristics and associated signs may be of help, and particularly the suspicion of neurocysticercosis should be raised in patients with late-onset seizures (even more in those with a normal neurological examination) and patients with a history of palpable nodes in the subcutaneous tissues or in the muscles. Interesting trends in gender and age distribution can be seen in clinical cases. A single enhancing brain lesion and basal subarachnoid neurocysticercosis seem to be more frequent in women, while multiple viable cysts seem more frequent in men. As it could be expected, calcifications are seen in older patients compared to viable cysts, and subarachnoid disease in even older patients. A single cysticercus granuloma is seen in younger patients, more markedly in India (Chandy et al. 1991; Rajshekhar and Chandy 2000). Albeit massive infections are in general terms rare, these are proportionally more frequent at younger ages (likely because these infections are clinically eloquent), and massive infections with diffuse inflammation (encephalitic cysticercosis) are most often seen in young women (Rangel et al. 1987).

4.2.4 Geographical Variations in the Pattern of Clinical Expression of Cysticercosis

Two trends can be defined from the literature. In the Indian subcontinent, the vast majority of clinical cases of neurocysticercosis occur in older children and young teenagers and correspond to one, less frequently two, already degenerating cysts (Chandy et al. 1991; Rajshekhar and Chandy 2000). This type of disease (solitary cysticercal granuloma, single enhancing lesion) has a particularly mild clinical expression and good prognosis. Also, cases with subcutaneous nodules are more frequently reported from Africa and Asia than from Latin America. Whether this results from the genetic diversity of the host inflammatory reaction to the parasite, or simply to different patterns of infection or infection burden, cannot be discerned from the existing information.

References

Barinagarrementeria F, Cantu C (2002) Cerebrovascular manifestations of neurocysticercosis. In: Singh G, Prabhakar S (eds) *Taenia solium* cysticercosis. From basic to clinical science. CABI Publishing, Oxon

Briceño CE, Biagi F, Martínez B (1961) Cisticercosis. Observaciones sobre 97 casos de autopsia. Prensa Med Mex 26:193–197

Chandy MJ, Rajshekhar V, Ghosh S, Prakash S, Joseph T, Abraham J, Chandi SM (1991) Single small enhancing CT lesions in Indian patients with epilepsy: clinical, radiological and pathological considerations. J Neurol Neurosurg Psychiatry 54:702–705

Chopra JS, Kaur U, Mahajan RC (1981) Cysticerciasis and epilepsy: a clinical and serological study. Trans R Soc Trop Med Hyg 75:518-520

Coker-Van MR, Subianto DB, Brown P, Diwan AR, Desowitz R, Garruto RM, Gibbs CJ Jr, Gajdusek DC (1981) ELISA antibodies to cysticerci of *Taenia solium* in human populations in New Guinea, Oceania, and Southeast Asia. Southeast Asian J Trop Med Public Health 12:499–505

Cruz ME, Schantz PM, Cruz I, Espinosa P, Preux PM, Cruz A, Benitez W, Tsang VC, Fermoso J, Dumas M (1999) Epilepsy and neurocysticercosis in an Andean community. Int J Epidemiol 28:799–803

Cysticercosis Working Group in Peru (1993) The marketing of cysticercotic pigs in the Sierra of Peru. Bull WHO 71:223–228

Del Brutto OH (1992) Cysticercosis and cerebrovascular disease: a review. J Neurol Neurosurg Psychiatry 55:252–254

Del Brutto OH (2012) Neurocysticercosis in a 2-year-old boy infected at home. Pathog Global Health 106:122–123

Del Brutto OH, Granados G, Talamas O, Sotelo J, Gorodezky C (1991) Genetic pattern of the HLA system: HLA A, B, C, DR, and DQ antigens in Mexican patients with parenchymal brain cysticercosis. Hum Biol 63:85–93

Del Brutto OH, Santibañez R, Noboa CA, Aguirre R, Díaz E, Alarcón TA (1992) Epilepsy due to neurocysticercosis: analysis of 203 patients. Neurology 42:389–392

Del Brutto OH, Santibánez R, Idrovo L, Rodriguez S, Diaz-Calderon E, Navas C, Gilman RH, Cuesta F, Mosquera A, González AE, Tsang VC, García HH (2005) Epilepsy and neurocysticercosis in Atahualpa: a door-to-door survey in rural coastal Ecuador. Epilepsia 46: 583–587

Diaz F, García HH, Gilman RH, Gonzales AE, Castro M, Tsang VC, Pilcher JB, Vasquez LE, Lescano M, Carcamo C, Madico G, Miranda E (1992) Epidemiology of taeniasis and cysticercosis in a Peruvian village. The Cysticercosis Working Group in Peru. Am J Epidemiol 135:875–882

Dorny P, Kabwe C, Kirezi K, Lukanu K, Lutumba P, Maketa V, Matondo P, Polman K, Praet N, Speybroeck N, Sumbu J (2012) Cysticercosis in the democratic Republic of Congo. Onderstepoort J Vet Res 79:E1

Escalante S (1977) Epidemiología de la cysticercosis en Perú. Rev Neuropsiquiatría 40:29–39

Fleury A, Gomez T, Alvarez I, Meza D, Huerta M, Chavarria A, Carrillo Mezo RA, Lloyd C, Dessein A, Preux PM, Dumas M, Larralde C, Sciutto E, Fragoso G (2003) High prevalence of calcified silent neurocysticercosis in a rural village of Mexico. Neuroepidemiology 22:139–145

Fleury A, Carrillo-Mezo R, Flisser A, Sciutto E, Corona T (2011) Subarachnoid basal neurocysticercosis: a focus on the most severe form of the disease. Expert Rev Anti Infect Ther 9:123–133

García HH, Del Brutto OH (2005) Neurocysticercosis: updated concepts about an old disease. Lancet Neurol 4:653–661

García HH, Martinez M, Gilman R, Herrera G, Tsang VCW, Pilcher JB, Díaz F, Verastegui M, Gallo C, Porras M, Alvarado M, Naranjo J, Miranda E (1991) Diagnosis of cysticercosis in endemic regions. Lancet 338:549–551

García HH, Gilman RH, Tovar MA, Flores E, Jo R, Tsang VC, Diaz F, Torres P, Miranda E (1995) Factors associated with *Taenia solium* cysticercosis: analysis of nine hundred forty-six Peruvian neurologic patients. Cysticercosis Working Group in Peru (CWG). Am J Trop Med Hyg 52:145–148

García HH, Gilman RH, Catacora M, Verastegui M, Gonzalez AE, Tsang VC (1997) Serologic evolution of neurocysticercosis patients after antiparasitic therapy. J Infect Dis 175:486–489

García HH, Talley A, Gilman RH, Zorrilla L, Pretell J (1999) Epilepsy and neurocysticercosis in a village in Huaraz, Peru. Clin Neurol Neurosurg 101:225–228

García HH, Gonzalez AE, Gilman RH, Palacios LG, Jimenez I, Rodriguez S, Verastegui M, Wilkins P, Tsang VC (2001) Transient antibody response in *Taenia solium* infection in field conditions-a major contributor to high seroprevalence. Am J Trop Med Hyg 65:31–32

García HH, Rodriguez S, Gilman RH, Gonzalez AE, Tsang VC (2012) Neurocysticercosis: is serology useful in the absence of brain imaging? Trop Med Int Health 17:1014–1018

Garcia-Noval J, Allan JC, Fletes C, Moreno E, DeMata F, Torres-Alvarez R, Soto de Alfaro H, Yurrita P, Higueros-Morales H, Mencos F, Craig PS (1996) Epidemiology of *Taenia solium* taeniasis and cysticercosis in two rural Guatemalan communities. Am J Trop Med Hyg 55:282–289

Garcia-Noval J, Moreno E, de Mata F, Soto de Alfaro H, Fletes C, Craig PS, Allan JC (2001) An epidemiological study of epilepsy and epileptic seizures in two rural Guatemalan communities. Ann Trop Med Parasitol 95:167–175

Gilman R, García HH, González AE, Dunleavy M, Verastegui M, Evans C (1999) Short cuts to development: methods to control the transmission of cysticercosis in developing countries. In: García HH, Martinez SM (eds) *Taenia solium* taeniasis/cysticercosis, 2nd edn. Editorial Universo, Lima

Gilman RH, Del Brutto OH, García HH, Martinez M (2000) Prevalence of taeniosis among patients with neurocysticercosis is related to severity of infection. Neurology 55:1062

Lescano AG, García HH, Gilman RH, Guezala MC, Tsang VC, Gavidia CM, Rodriguez S, Moulton LH, Green JA, Gonzalez AE (2007) Swine cysticercosis hotspots surrounding *Taenia solium* tapeworm carriers. Am J Trop Med Hyg 76:376–383

Lescano AG, García HH, Gilman RH, Gavidia CM, Tsang VC, Rodriguez S, Moulton LH, Villaran MV, Montano SM, Gonzalez AE (2009) *Taenia solium* cysticercosis hotspots surrounding tapeworm carriers: clustering on human seroprevalence but not on seizures. PLoS Negl Trop Dis 3:e371

Lino-Junior Rde S, Faleiros AC, Vinaud MC, Oliveira FA, Guimaraes JV, Reis MA, Texeira Vde P (2007) Anatomopathological aspects of neurocysticercosis in autopsied patients. Arq Neuropsiquiatr 65:87–91

Lobato RD, Lamas E, Portillo JM, Roger R, Esparza J, Rivas JJ, Muñoz MJ (1981) Hydrocephalus in cerebral cysticercosis. Pathogenic and therapeutic considerations. J Neurosurg 55:786–793

Medina MT, Rosas E, Rubio-Donnadieu F, Sotelo J (1990) Neurocysticercosis as the main cause of late-onset epilepsy in Mexico. Arch Intern Med 150:325–327

Medina MT, Durón RM, Martínez L, Osorio JR, Estrada AL, Zúniga C, Cartagena D, Collins JS, Holden KR (2005) Prevalence, incidence, and etiology of epilepsies in rural Honduras: the Salamá Study. Epilepsia 46:124–131

Meza-Lucas A, Carmona-Miranda L, Garcia-Jeronimo RC, Torrero-Miranda A, Gonzalez-Hidalgo G, López-Castellanos G, Correa D (2003) Short report: limited and short-lasting humoral response in *Taenia solium*: seropositive households compared with patients with neurocysticercosis. Am J Trop Med Hyg 69:223–227

Michel P, Callies P, Raharison H, Guyon P, Holvoet L, Genin C (1993) Epidémiologie de la cysticercose a Madagascar. Bull Soc Path Ex 8:62–67

Montano SM, Villaran MV, Ylquimiche L, Figueroa JJ, Rodriguez S, Bautista CT, Gonzalez AE, Tsang VC, Gilman RH, García HH (2005) Neurocysticercosis: association between seizures, serology, and brain CT in rural Peru. Neurology 65:229–233

Mwanjali G, Kihamia C, Kakoko DVC, Lekule F, Ngowi H, Johansen MV, Thamsborg SM, Willingham AL 3rd (2013) Prevalence and risk factors associated with human *Taenia solium* infections in Mbozi District, Mbeya Region, Tanzania. PLoS Negl Trop Dis 7:e2102

Mwape KE, Phiri IK, Praet N, Speybroeck N, Muma JB, Dorny P, Gabriel S (2013) The incidence of human cysticercosis in a rural community of eastern Zambia. PLoS Negl Trop Dis 7:e2142

Ndimubanzi PC, Carabin H, Budke CM, Nguyen H, Qian YJ, Rainwater E, Dickey M, Reynolds S, Stoner JA (2010) A systematic review of the frequency of neurocysticercosis with a focus on people with epilepsy. PLoS Negl Trop Dis 4:e870

Nguekam JP, Zoli AP, Zogo PO, Kamga AC, Speybroeck N, Dorny P, Brandt J, Losson B, Geerts S (2003) A seroepidemiological study of human cysticercosis in West Cameroon. Trop Med Int Health 8:144–149

O'Neal SE, Moyano LM, Ayvar V, Gonzalvez G, Diaz A, Rodriguez S, Wilkins PP, Tsang VC, Gilman RH, García HH, Gonzalez AE (2012) Geographic correlation between tapeworm carriers and heavily infected cysticercotic pigs. PLoS Negl Trop Dis 6:e1953

Praet N, Speybroeck N, Rodriguez-Hidalgo R, Benitez-Ortiz W, Berkvens D, Brandt J, Saegerman C, Dorny P (2010) Age-related infection and transmission patterns of human cysticercosis. Int J Parasitol 40:85–90

Prasad KN, Verma A, Srivastava S, Gupta RK, Pandey CM, Paliwal VK (2011) An epidemiological study of asymptomatic neurocysticercosis in a pig farming community in northern India. Trans R Soc Trop Med Hyg 105:531–536

Preux PM, Druet-Cabanac M (2005) Epidemiology and aetiology of epilepsy in sub-Saharan Africa. Lancet Neurol 4:21–31

Rajshekhar V, Chandy MJ (eds) (2000) Solitary cysticercus granuloma: the disappearing lesion. Orient Longman Limited, Chennai

Rangel R, Torres B, Del Brutto O, Sotelo J (1987) Cysticercotic encephalitis: a severe form in young females. Am J Trop Med Hyg 36:387–392

Rosas N, Sotelo J, Nieto D (1986) ELISA in the diagnosis of neurocysticercosis. Arch Neurol 43: 353–356

Sánchez AL, Lindback J, Schantz PM, Sone M, Sakai H, Medina MT, Ljungstrom I (1999) A population-based, case–control study of *Taenia solium* taeniasis and cysticercosis. Ann Trop Med Parasitol 93:247–258

Sarti E, Schantz PM, Plancarte A, Wilson M, Gutierrez IO, Lopez AS, Roberts J, Flisser A (1992) Prevalence and risk factors for *Taenia solium* taeniasis and cysticercosis in humans and pigs in a village in Morelos, Mexico. Am J Trop Med Hyg 46:677–685

Sarti E, Schantz PM, Plancarte A, Wilson M, Gutierrez OI, Aguilera J, Roberts J, Flisser A (1994) Epidemiological investigation of *Taenia solium* taeniasis and cysticercosis in a rural village of Michoacan State, Mexico. Trans R Soc Trop Med Hyg 88:49–52

Sarti-Gutierrez EJ, Schantz PM, Lara-Aguilera R, Gomez Dandoy H, Flisser A (1988) T*aenia solium* taeniasis and cysticercosis in a Mexican village. Trop Med Parasitol 39:194–198

Schantz PM, Sarti E, Plancarte A, Wilson M, Criales JL, Roberts J, Flisser A (1994) Community-based epidemiological investigations of cysticercosis due to *Taenia solium*: comparison of serological screening tests and clinical findings in two populations in Mexico. Clin Infect Dis 18:879–885

Schenone H (1982) Epidemiology of human cysticercosis in Latin America. In: Flisser A, Willms K, Laclette JP, Larralde C, Ridaura C, Beltran F (eds) Cysticercosis: present state of knowledge and perspectives. Academic, New York

Suástegui-Roman RA, Soto-Hernandez JL, Sotelo J (1996) Effects of prednisone on ventriculoperitoneal shunt function in hydrocephalus secondary to cysticercosis: a preliminary study. J Neurosurg 84:629–633

Verma A, Prasad KN, Gupta RK, Singh AK, Nyati KK, Rizwan A, Pandey CM, Paliwal VK (2010) Toll-like receptor 4 polymorphism and its association with symptomatic neurocysticercosis. J Infect Dis 202:1219–1225

Wilson M, Bryan RT, Fried JA, Ware DA, Schantz PM, Pilcher JB, Tsang VC (1991) Clinical evaluation of the cysticercosis enzyme-linked immunoelectrotransfer blot in patients with neurocysticercosis. J Infect Dis 164:1007–1009

Winkler AS, Blocher J, Auer H, Gotwald T, Matuja W, Schmutzhard E (2009) Epilepsy and neurocysticercosis in rural Tanzania-an imaging study. Epilepsia 50:987–993

Winkler AS, Schmutzhard E, Willingham AL (2010) Epilepsy in Sub-Saharan Africa: focus on neurocysticercosis. Res Adv Epilepsy 1:1–30

Woodhouse E, Flisser A, Larralde C (1982) Seroepidemiology of human cysticercosis in Mexico. In: Flisser A, Willms K, Laclette JP, Larralde C, Ridaura C, Beltran F (eds) Cysticercosis: present state of knowledge and perspectives. Academic, New York

5 Epidemiology of Human Cysticercosis in Non-endemic Regions and in the Traveler

Non-endemic areas for cysticercosis could be defined as those places where the life cycle of *Taenia solium* cannot be fully completed, as one or more of the several interrelated steps needed for its completion are not present (Del Brutto and García 2012). Those steps include the following: (1) the presence of *Taenia* carriers harboring the adult tapeworm in the intestine, (2) the practice of open-air fecalism or improper disposal of human feces, (3) the allowance of free-roaming pigs having access to human feces, and (4) the consumption of undercooked pork. When these four steps are associated with illiteracy, poverty, poor sanitization, and a warm climate, the prevalence of human taeniasis as well as that of human and porcine cysticercosis may attain endemic consequences and may represent a major public health problem (García et al. 2007). As noted in the preceding chapter, this is what happens in most of the developing world.

In contrast, industrialized nations of North America, Western Europe, and Oceania have developed, through the past decades, secure systems of sewage disposal and adequate mechanisms of husbandry, precluding pigs to be in contact with human feces, thus avoiding the life cycle of *Taenia solium* to be completed. There are also some regions—developed or not—where the life cycle of the tapeworm cannot be completed, as their religious beliefs prohibit the consumption of pork; these include Israel and Muslim countries of the Arab world.

Traditionally considered non-endemic regions have been faced, during the past decades, with the massive immigration of people coming from cysticercosis-endemic areas that, together with increased overseas travelling and refugee movements, progressively increased the number of patients with neurocysticercosis in these regions. As known, the first non-endemic country experiencing such increased prevalence was the UK after the massive return of soldiers on duty from the Indian subcontinent (MacArthur 1934). This has been followed by a number of bouts of human cysticercosis in other non-endemic regions, related to either mass movement of people or infected swine from endemic to non-endemic areas (Fig. 5.1).

O.H. Del Brutto, H.H. García, *Cysticercosis of the Human Nervous System*,
DOI 10.1007/978-3-642-39022-7_5,

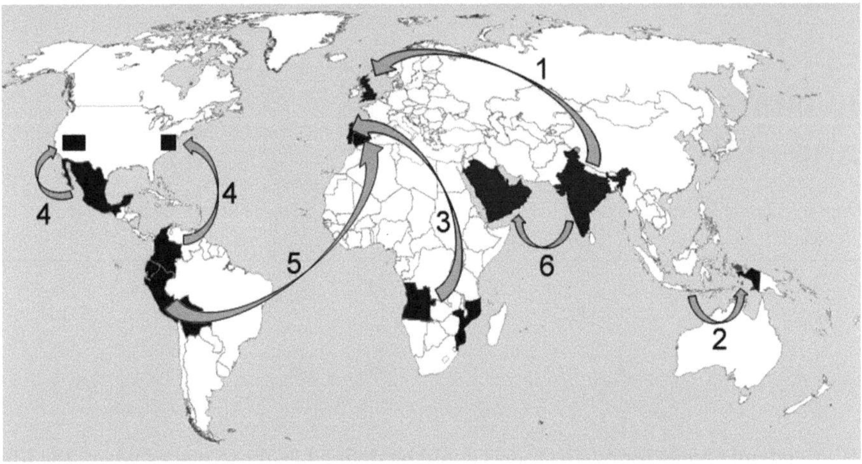

Fig. 5.1 World map showing major outbreaks of human cysticercosis related to mass movement of people or infected swine from endemic to non-endemic areas. (*1*) Return of British soldiers from India to England; (*2*) gift of infected swine from Bali to Irian Jaya; (*3*) mass return of Portuguese living in African colonies after liberty wars in Angola and Mozambique; (*4*) migratory movements of people from Mexico and South America to the USA (mainly to the Southwestern USA and the New York city area); (*5*) mass migration of people from Ecuador, Perú and Bolivia to Spain; (*6*) migration of people from India to countries of the Arabian Peninsula (Reproduced with permission from: Del Brutto (2012d))

While initially almost all cases of human cysticercosis diagnosed in non-endemic regions occurred in immigrants or in people repatriated from disease-endemic areas, the problem has been complicated by the increasing occurrence of autochthonous cases, i.e., patients born in non-endemic regions who have never been abroad (García 2012). This created confusion among physicians and health authorities who, mostly unaware on the mechanisms of disease transmission, could not found an explanation to this. It was later understood that, in the same way as people with cysticercosis migrate from disease-endemic to non-endemic areas, asymptomatic *Taenia* carriers also migrate, and they represent the source of infection to local people, thus increasing and favoring the spread of cysticercosis in non-endemic countries without the need of infected pigs (Schantz et al. 1992). Despite this evidence, little has been done to detect the *Taenia* carrier among close contacts of neurocysticercosis patients living in non-endemic regions. Moreover, the illegal status of many of these persons complicates even more the access to household contacts of neurocysticercosis patients in the search of *Taenia* carriers.

Nowadays, there is almost no country, considered as non-endemic, where at least a few human cases of cysticercosis have been reported. In general terms, it can be stated that more than 80 % of these cases occur in immigrants from disease-endemic areas. However, on the basis of disease expression of neurocysticercosis in some of these immigrants, it is possible that they were not infected back in their country of origin (before migration), but after they arrived to the non-endemic area (Del Brutto et al. 2012a, b).

5.1 The United States

Human cysticercosis was considered exceedingly rare in the USA, with only 42 cases being reported since 1857–1954 (Campagna et al. 1954). During the following years and up to the late 1970s, a number of case reports and small series of patients also called the attention to the occurrence of this parasitic disease in the USA, suggesting that its prevalence was on the rise (Bentson et al. 1977; Carmalt et al. 1975; Cohen 1962; DeFeo et al. 1975; White et al. 1957). Together with the mass immigration of people from Mexico and other Latin American countries, the first bouts of human cysticercosis were noticed in the Southwestern USA—mainly in the states of California and Texas—where more than 500 patients were recognized during the 1980s, most of whom were of Mexican origin (Earnest et al. 1987; Grisolia and Wiederholt 1982; Loo and Braude 1982; McCormick 1985; McCormick et al. 1982; Mitchell and Snodgrass 1985; Richards et al. 1985). A literature review of case series of neurocysticercosis patients diagnosed in the USA from 1980 to 2004 found almost 1,500 patients (Wallin and Kurtzke 2004), a number that increased to more than 5,000 when more recent publications were included (Croker et al. 2010, 2012; O'Neal et al. 2011; Serpa et al. 2011). However, these alarming numbers are clearly an underestimate since neurocysticercosis is not a reportable disease in most states (Serpa and White 2012). Assessing the actual burden of neurocysticercosis in the USA is complicated. Conservative analysis estimates annual incidence rates ranging from 1.5 to 5.8 cases per 100,000 in Hispanics and from 0.02 to 0.5 cases per 100,000 in non-Hispanic whites (O'Neal et al. 2011; Sorvillo et al. 1992; Townes et al. 2004).

As noted, most cases of neurocysticercosis in the USA occur among immigrants from disease-endemic countries. However, up to 7 % of cases in some series have been autochthonous, reflecting the occurrence of locally acquired disease (Sorvillo et al. 1992). A literature review showed 78 well-documented cases of neurocysticercosis acquired in the USA up to the year 2005; in 21 % of these patients, a close contact infected with the adult *Taenia solium* was found (Sorvillo et al. 2011). The most precisely described example of this form of acquisition of human disease in the USA was reported in the Orthodox Jewish community of New York City, where some of their members (child and adults) developed neurocysticercosis as the result of infection from domestic employees recently emigrated from Latin America who were found to be *Taenia solium* carriers, infecting people for whom they worked through non-hygienic handling of food (Schantz et al. 1992).

While most patients with neurocysticercosis diagnosed in the USA presented with seizures and parenchymal brain granulomas, thus having a benign prognosis, some developed severe forms of the disease such as giant subarachnoid cysts or hydrocephalus (Kelesidis and Tsiodras 2011) and others have even died as the result of the infection (Sorvillo et al. 2007).

5.2 Canada

Swine cysticercosis has been recognized in Canada since the nineteenth century (Cameron 1934). However, according to Canadian health authorities, there were no reported cases of human cysticercosis in the country up to 1956 (White et al. 1957). During the 1970s, isolated case reports called the attention to the occurrence of this condition among immigrants to Canada (Ali-Khan et al. 1979; Scholten et al. 1976), and by 1986, the first series of eight patients with neurocysticercosis was reported from Montreal; all of them were immigrants from disease-endemic areas (Leblanc et al. 1986). Soon thereafter, Sheth et al. (1998) reported 29 patients with neurocysticercosis evaluated at Toronto over 6 years; however, no details were given on the citizenship status of these patients. More recently, several reports have confirmed the sporadic occurrence of this parasitic disease in Canada, always in immigrants from endemic areas (Boulos et al. 2010; Brophy and Keystone 2006; Burneo and McLachlan 2005; Burneo et al. 2009; Hajek and Keystone 2009).

A recent review of the literature of human cysticercosis in Canada (Del Brutto 2012a) identified a total of 55 patients with neurocysticercosis, most of who were diagnosed over the past two decades. Patients were mainly concentrated in Ontario and Quebec. Information on the citizenship status was available in only 28 of these patients. Twenty-seven of them were immigrants from disease-endemic areas, and the remaining case was a Canadian citizen with history of traveling abroad. Information on the time elapsed between immigration and symptomatic disease varied widely from a few months to more than 30 years, raising the question of whether some of these immigrants could have been infected while already living in Canada. While literature information is rather incomplete, it could be inferred that most of these patients presented with seizures related to the occurrence of parenchymal brain cysts or granulomas. Stool examinations were performed in just a few patients, and some of them were positive for *Taenia* eggs. In no case were household contacts investigated for taeniasis. This recent review let it clear that human cysticercosis is an emerging problem in Canada. However, no actions have yet been taken to reduce its impact burden or to prevent further spread of the disease (Burneo 2012).

5.3 Western European Countries

While human cysticercosis has been linked to Western Europe for centuries, the disease became neglected during the past decades just because it was considered rare (Overbosch et al. 2002). A re-emergency of the disease is currently happening in some of these countries due to the massive immigration of people from Africa and South America that took place during the past two decades. For example, neurocysticercosis

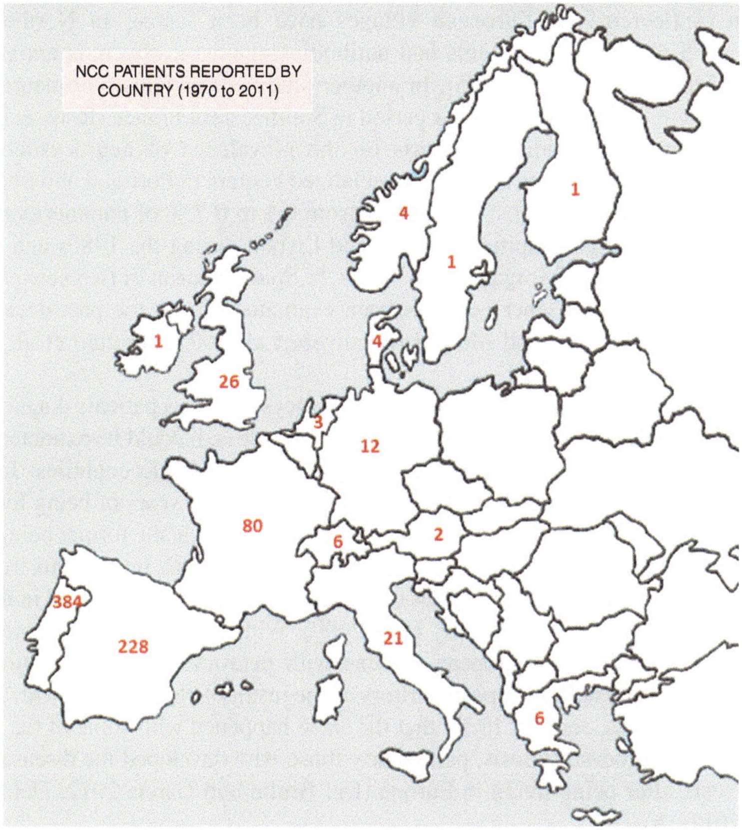

Fig. 5.2 Map of Western Europe showing countries with the number of reported cases of neurocysticercosis from 1970 to 2011

in Spain remained endemic during the twentieth century up to the 1980s, where its prevalence was reduced to 4.3 % per 100,000 inhabitants, with cases confined to rural areas (García-Albea 1991). Together with the growing number of immigrants, more than 100 "urban" cases have been recognized during the past decade, mostly among South American immigrants (Giménez-Roldán et al. 2003; Más-Sesé et al. 2008; Ramos et al. 2011; Ruiz et al. 2011). Likewise, most of the cases reported from Portugal were temporarily related to the massive return of Portuguese citizens from their African colonies after their independence (Almeida-Pinto et al. 1988; Monteiro 1993; Monteiro et al. 1986, 1992, 1993, 1995; Morgado et al. 1994).

A recent literature search identified 779 patients with neurocysticercosis diagnosed in Western Europe from 1970 to 2011 (Del Brutto 2012b). Countries with more reported patients were Portugal ($n=384$), Spain ($n=228$), and France ($n=80$), followed by the UK, Italy, and Germany; in the remaining countries, there was either no data or the reported number of patients was five or less (Fig. 5.2). Community-based studies attempting to determine the prevalence of human cysticercosis is European villages have been scarce. In Northeastern Portugal, 5 % of the inhabitants had antibodies against cysticercus antigens in serum (Meneses-Monteiro 1995). In another study, 29 patients with neurocysticercosis were found over a 10-year period in Southeastern France (Rousseau et al. 1999). Some information also exists on the prevalence of neurocysticercosis among neurologic patients attending specialized centers in Portugal and Spain. In the former, neurocysticercosis accounted from 0.1 to 0.7 % of patients evaluated with CT at two large hospitals in Porto and Lisbon during the 1980s and 1990s (Monteiro et al. 1992; Morgado et al. 1994). In Spain, a mean of two new patients with neurocysticercosis per year has been evaluated during the past decades at large hospitals in Vigo and Madrid (Esquivel et al. 2005; Fandiño et al. 1989; García-Albea 1989).

Information on the citizenship status of neurocysticercosis patients diagnosed in Western European countries is incomplete. Nevertheless, it could be extracted from the literature that most are immigrants from disease-endemic countries. In such cases, symptomatic disease has occurred after a mean of 5 years of being living in Europe. Most patients came from South America and Africa, the former being most often diagnosed in Spain and the latter in France and Portugal. Immigrants from the Indian subcontinent and other regions of Asia were more often diagnosed in the UK (Kennedy and Schon 1991; Wadley et al. 2000). With the exception of some international travelers, most European citizens with neurocysticercosis acquired the infection in the Western European territory as the result of close contact with *Taenia* carrier immigrants, and it is likely that the same happened with some of the immigrants with neurocysticercosis, particularly those who developed the disease 20 or more years after being living in Europe (Del Brutto and García 2012; Del Brutto et al. 2012a, b).

5.4 Countries of the Arab World

Due to religious laws prohibiting swine breeding and consumption of pork, human cysticercosis has been considered inexistent in Muslin countries of the Arab world with the exception of sporadic cases seen in immigrants from India during the last two decades of the twentieth century (Chandy et al. 1989). More recently, a number of case series suggested that the prevalence of neurocysticercosis in the region is on the rise. Indeed, a literature search from the year 2000 to 2011 allowed the identification of 39 patients with neurocysticercosis (Del Brutto 2013). Most of these patients were from Kuwait, although some came from Qatar and Saudi Arabia. Patients were most often young women, and most had a benign clinical course, since the most common pattern of disease expression of neurocysticercosis in Arabian is a single parenchymal brain cysticercal granuloma (Abdulla et al. 2006; Al Shahrani et al. 2003; Al-Khuwaitir et al. 2005; Hamed and El-Metaal 2007; Hira et al. 2004; Hussein et al. 2003; Khan et al. 2011).

Most patients reported from the Arab world were autochthonous, and occurred in the context of wealthy families employing babysitters and housekeepers from disease-endemic areas. Indeed, familiar cases of neurocysticercosis were recorded in some cases. Most important, stool examinations in household contacts of some of these patients allowed the identification of *Taenia* carriers.

5.5 Australia and New Zealand

Taeniasis and cysticercosis have always been considered rare in Oceania. However, a number of recent reports suggest that neurocysticercosis is increasingly recognized in this Continent. A total of 39 neurocysticercosis patients have been reported from Australia, 33 of whom were published in the past two decades. This does not seem to be the case of New Zealand, as only two patients have been described in that country (Pybus and Heron 2012; Wickremesekera et al. 1996).

More than 75 % of Australian patients with neurocysticercosis were immigrants from disease- endemic areas, and the remaining were Australian citizens who had history of traveling to endemic regions (Crimmins et al. 1990; Lucey et al. 2010; Hewagama et al. 2010; Walker et al. 1991; Yong and Warren 1994). So, it seems that Australia is free of locally acquired disease. However, it is also possible that some immigrants who developed neurocysticercosis while living in Australia were not infected abroad, ahypothesis that is supported by the fact that some of these persons have developed the disease more than 10 years after they had migrated from their countries of origin (Del Brutto 2012b). Unfortunately, case reports and case series of neurocysticercosis patients diagnosed in Australia have not included information on whether household contacts of diagnosed patients had been investigated for the presence of *Taenia* carriers.

5.6 International Travelers

Information on neurocysticercosis in international travelers from non-endemic to disease-endemic countries was not well characterized until a recent literature search identified a total of 52 such cases over the past three decades, a very low number considering the millions of people who have traveled from non-endemic to disease-endemic regions during those years (Del Brutto 2012c). However, it is likely that such review just captured the tip of the iceberg, as many undiagnosed and unreported cases may exist. Most of the reported patients were middle-aged adults originally from England, Australia, Israel, Japan and France, and common places of destination were the Indian subcontinent, South East Asia, Latin America and Africa. Only two of these patients had history of short-term sojourns (up to 3 months). The time elapsed between return of the travel to the appearance of symptoms varied, but most patients became symptomatic at least 2 years after returning home. Most patients had seizures as the primary or sole manifestation of the diseases, and neuroimaging studies most often showed a single cysticercus granuloma located in the brain parenchyma. Also, the prognosis was good in most cases.

To acquire the disease, travelers must be in contact with a *Taenia* carrier, who will infect them through non-hygienic handling of food. Another possibility is that travelers get in direct contact with human feces by visiting places where open-air defecation is a common practice. It is also possible that travelers first become *Taenia solium* carriers (by ingesting undercooked pork infected by cysticerci) and then infected themselves by the fecal-oral route. The most common pattern of neurocysticercosis expression in travelers, i.e., a single cysticercus granuloma, suggests that the usual form of disease acquisition is through sporadic contact with *Taenia* carrier food handlers. Otherwise, travelers would more often present with heavier infections, which are observed in *Taenia* carriers who infected themselves or in those who ingest a heavy load of *Taenia solium* eggs from nature (Gilman et al. 2000).

An unsolved issue is why a sizable percentage of travelers have developed symptoms several years after returning home. While it is true that cysticerci may live for years within the nervous system, there are some forms of the disease that most often represent "recent" infections producing symptoms in the very few months after disease acquisition. This is the case of the single cysticercal granuloma, a particular form of neurocysticercosis in which the host immune system actively reacts to the implantation of the metacestode of *Taenia solium* in the brain parenchyma (Del Brutto et al. 2012a, b). Since most travelers had this form of the disease, one would expect symptoms to occur soon after returning home. So, it is possible that what was presented on neuroimaging studies performed at the time of symptoms (seizures) were not active cysticercal granulomas, but the late sequelae of an infection that was previously handled by the host immune system without producing symptoms. As will be described in subsequent chapters of this book, recent evidence suggests that calcifications may experience periodic morphological changes that may expose parasitic antigenic material to the host, causing transient inflammatory changes in the brain parenchyma that may be the cause of seizures and changes on neuroimaging studies—brain swelling and ring-enhancing appearance of the lesion—resembling those seen in patients with acute cysticercal granulomas (Nash et al. 2004).

References

Abdulla JK, Al-Tawari AA, Cindro-Heberle L, Gopinath MS, Neubauer D (2006) Neurocysticercosis in non-endemic Muslin areas: a report of seven cases from Kuwait. J Pediatr Neurol 4:257–260

Al Shahrani D, Frayha HH, Dabbagh O, Al Shail E (2003) First case of neurocysticercosis in Saudi Arabia. J Trop Pediatr 49:58–60

Al-Khuwaitir TS, Al-Moghairi AM, El Zain FN, Al-Zayer WS (2005) Neurocysticercosis in central Saudi Arabia. Neuroscience 10:226–229

Ali-Khan Z, Chayasirisobhon S, Aubé M (1979) Human cysticercosis: a probable case of cerebral cysticercosis with generalized subcutaneous nodular lesions. Can J Neurol Sci 6:371–374

Almeida-Pinto J, Veiga-Pires JA, Stocker A, Coelho T, Monteiro L (1988) Cysticercosis of the brain. The value of computed tomography. Acta Radiol 29:625–628

Bentson JR, Wilson GH, Helmer E, Winter J (1977) Computed tomography in intracranial cysticercosis. J Comput Assist Tomogr 1:464–471

Boulos MI, Aviv RI, Lee L (2010) Spinal neurocysticercosis manifesting as recurrent aseptic meningitis. Can J Neurol Sci 37:878–880
Brophy J, Keystone J (2006) First seizure in a new immigrant. Paediatr Child Health 11:345–347
Burneo JG (2012) Neurocysticercosis: a foreign parasite looking for "permanent resident" status? Can J Neurol Sci 39:269–270
Burneo JG, McLachlan RS (2005) Neurocysticercosis in Canada: still a rare disease? Can J Neurol Sci 32(Suppl 1):S67
Burneo J, Plener I, García HH (2009) Neurocysticercosis in a patient in Canada. Can Med Assoc J 180:639–642
Cameron TWM (1934) Sir William Osler—parasitologist. Can Med Assoc J 30:553–556
Campagna M, Swartzwelder C, Comer EO (1954) Human cysticercosis in the United States. J Parasitol 40(Suppl):46 (abstract)
Carmalt JE, Theis J, Goldstein E (1975) Spinal cysticercosis. West J Med 123:311–313
Chandy MJ, Sharma RR, Lad SD, Manchanda A (1989) Focal epilepsy and a single small enhancing cortical ring lesion on computed tomography in Asians. Emirates Med J 7:190–197
Cohen B (1962) Cysticercosis cerebri: a case report. South Med J 55:48–55
Crimmins D, Collignon PJ, Dwyer D, Danta G (1990) Neurocysticercosis: an under-recognized cause of neurological problems. Med J Aust 152:434–438
Croker C, Reporter R, Mascola L (2010) Use of statewide hospital discharge data to evaluate the economic burden of neurocysticercosis in Los Angeles County (1991-2008). Am J Trop Med Hyg 83:106–110
Croker C, Redelings M, Reporter R, Sorvillo F, Mascola L, Wilkins P (2012) The impact of neurocysticercosis in California: a review of hospitalized cases. PLoS Neglected Trop Dis 6:e1480
DeFeo D, Foltz EL, Hamilton AE (1975) Double compartment hydrocephalus in a patient with cysticercotic meningitis. Surg Neurol 4:247–251
Del Brutto OH (2012a) A review of cases of human cysticercosis in Canada. Can J Neurol Sci 39:319–322
Del Brutto OH (2012b) Neurocysticercosis in Australia: still free of autochthonous cases? Med J Aust 196:385
Del Brutto OH (2012c) Neurocysticercosis among international travelers to disease-endemic areas. J Travel Med 19:112–117
Del Brutto OH (2012d) Neurocysticercosis. Continuum (Minneap Minn) 18:1392–1416
Del Brutto OH (2013) Neurocysticercosis in the Arabian Peninsula. Emerg Infect Dis 19(1): 172–174
Del Brutto OH, García HH (2012) Neurocysticercosis in nonendemic countries: time for a reappraisal. Neuroepidemiology 39:145–146
Del Brutto OH, Nash TE, García HH (2012a) Cysticerci-related single parenchymal brain enhancing lesions in non-endemic countries. J Neurol Sci 319:32–36
Del Brutto VJ, Del Brutto OH, Ochoa E, García HH (2012b) Single parenchymal brain cysticercus: relationship between age of patients and evolutive stage of parasites. Neurol Res 34: 967–970
Earnest MP, Reller B, Filley CM, Grek AJ (1987) Neurocysticercosis in the United States: 35 cases and a review. Rev Infect Dis 9:961–979
Esquivel A, Díaz-Otero F, Giménez-Roldán S (2005) Growing frequency of neurocysticercosis in Madrid (Spain). Neurologia 20:116–120
Fandiño J, Rodriguez M, Pastor A, Viladrich A, Botana C, Gómez-Bueno J (1989) Cysticercose cérébrale. Dix cases Rev Neurol (Paris) 145:389–392
García HH (2012) Neurocysticercosis in immigrant populations. J Travel Med 19:73–75
García HH, Gonzalez AE, Del Brutto OH, Tsang VC, Llanos-Zavalaga F, Gonzalvez G, Romero J, Gilman RH (2007) Strategies for the elimination of taeniasis/cysticercosis. J Neurol Sci 262:153–157
García-Albea E (1989) Cisticercosis en España. Algunos datos epidemiológicos. Rev Clin Esp 184:3–6

García-Albea E (1991) Cisticercosis cerebral: aportaciones al conocimiento de una enfermedad endémica en España e Hispanoamérica. AránEdiciones S.A, Madrid
Gilman RH, Del Brutto OH, García HH, Martinez M (2000) Prevalence of taeniosis among patients with neurocysticercosis is related to severity of infection. Neurology 55:1062
Giménez-Roldán S, Díaz F, Esquivel A (2003) Neurocisticercosis e inmigración. Neurologia 18:385–388
Grisolia JS, Wiederholt WC (1982) CNS cysticercosis. Arch Neurol 39:540–544
Hajek J, Keystone J (2009) Intraventricular neurocysticercosis managed with albendazole and dexamethasone. Can J Neurol Sci 36:102–104
Hamed SA, El-Metaal HE (2007) Unusual presentations of neurocysticercosis. Acta Neurol Scand 115:192–198
Hewagama SS, Darby JD, Sheorey H, Daffy JR (2010) Seizures related to praziquantel therapy in neurocysticercosis. Med J Aust 193:246–247
Hira PR, Francis I, Abdella NA, Gupta R, Ai-Ali FM, Grover S, Khalid N, Abdeen S, Igbal J, Wilson M, Tsang VC (2004) Cysticercosis: imported and autochthonous infections in Kuwait. Trans R Soc Trop Med Hyg 98:233–239
Hussein FMY, Alhajri FA, Buriki KB, El Beltaji AH, Ovais MI, Almuhtaseb S (2003) Neurocysticercosis in Kuwait: computerized tomography and magnetic resonance imaging findings. Kuwait Med J 35:187–191
Kelesidis T, Tsiodras S (2011) Extraparenchymal neurocysticercosis in the United States. Am J Med Sci 344:79–82
Kennedy A, Schon F (1991) Epilepsy: disappearing lesions appearing in the United Kingdom. Br Med J 302:933–935
Khan FY, Iman YZ, Kamel H, Shafaee M (2011) Neurocysticercosis in Qatari patients: case reports. Trav Med Infect Dis 9:298–302
Leblanc R, Knowles KF, Melanson D, MacLean JD, Rouleau G, Farmer J-P (1986) Neurocysticercosis: surgical and medical management with praziquantel. Neurosurgery 18: 419–427
Loo L, Braude A (1982) Cerebral cysticercosis in San Diego. A report of 23 cases and a review of the literature. Medicine (Baltimore) 61:341–359
Lucey JM, McCarthy J, Burgner DP (2010) Encysted seizures: status epilepticus in a recently resettled refugee child. Med J Aust 192:237
MacArthur WP (1934) Cysticercosis as seen in the British Army with special reference to the production of epilepsy. Trans R Soc Trop Med Hyg 27:343–363
Más-Sesé G, Vives-Piñera I, Fernández-Barreiro A, Martínez-Lage JF, Martínez-Salcedo E, Alarcón-Martínez H, Domingo-Jiménez R, Puche-Mira A, Casas-Fernández C (2008) Estudio descriptivo de neurocisticercosis en un hospital terceario. Rev Neurol 46:194–196
McCormick GF (1985) Cysticercosis—review of 230 patients. Bull Clin Neurosci 50:76–101
McCormick GF, Zee C-S, Heiden J (1982) Cysticercosis cerebri. Review of 127 cases. Arch Neurol 39:534–539
Meneses-Monteiro LAS (1995) Neurocysticercosis in the North of Portugal. Arq Neurospichiat 53:3 (abstract)
Mitchell WG, Snodgrass SR (1985) Intraparenchymal cerebral cysticercosis in children: a benign prognosis. Pediatr Neurol 1:151–156
Monteiro L (1993) Neurocisticercose – uma parasitose (ainda) endémica no Norte de Portugal. Revista Portuguesa de Doençias Infecciosas 1:11–16
Monteiro L, Stocker A, Seca R (1986) TC, epilepsia e calcificacoescerebrais; análise de 73 casos. Bol Liga Port Epilepsia 2(Supp):47–53
Monteiro L, Coelho T, Stocker A (1992) Neurocysticercosis – a review of 231 cases. Infection 20:61–65
Monteiro L, Almeida-Pinto J, Stocker A, Sampaio-Silva M (1993) Active neurocysticercosis, parenchymal and extraparenchymal: a study of 38 patients. J Neurol 241:15–21
Monteiro L, Nunes B, Mendonca D, Lopes J (1995) Spectrum of epilepsy in neurocysticercosis: a long-term follow-up of 143 patients. Acta Neurol Scand 92:33–40

Morgado C, Gomes LB, de Campos JG (1994) Neurocysticercose. Análise imagiológica de 35 casos. Acta Med Port 7:269–275

Nash TE, Del Brutto OH, Butman JA, Corona T, Delgado-Escueta A, Duron RM, Evans CA, Gilman RH, Gonzalez AE, Loeb JA, Medina MT, Pietsch-Escueta S, Pretell EJ, Takayanagui OM, Theodore W, Tsang VC, García HH (2004) Calcific neurocysticercosis and epileptogenesis. Neurology 62:1934–1938

O'Neal S, Noh J, Wilkins P, Keene W, Lambert W, Anderson J, Luman JC, Townes J (2011) Taenia solium tapeworm infection, Oregon, 2006-2009. Emerg Infect Dis 17:1030–1036

Overbosch D, Oosterhuis JW, Kortbeek LM, Garcia-Albea E (2002) Neurocysticercosis in Europe. In: Craig P, Pawlowsky Z (eds) Cestode Zoonoses: echinococcosis and cysticercosis. Ios Press, Amsterdam

Pybus M, Heron P (2012) Neurocysticercosis: an unusual cause of seizure in the New Zealand setting. J Paediatr Child Health 48:E119–E121

Ramos JM, Masiá M, Padilla S, Escolano C, Bernal E, Gutiérrez F (2011) Enfermedades importadas y no importadas en la población inmigrante. Una década de experiencia desde una unidad de enfermedades infecciosas. Enf Infecc Microbiol Clin 29:185–192

Richards FO, Schantz PM, Ruiz-Tiben E, Sorvillo FJ (1985) Cysticercosis in Los Angeles County. JAMA 254:3444–3448

Rousseau MC, Guillotel B, Delmont J (1999) Neurocysticercose dans le Sud-Est de la France entre 1988 et 1998. Presse Med 28:2141–2144

Ruiz S, García-Vázquez E, Picazo R, Hernández A, Herrero JA, Gómez J (2011) La neurocisticercosis en Murcia. Rev Clin Esp 211:133–138

Schantz PM, Moore AC, Muñoz JL, Hartman BJ, Schaefer JA, Aron AM, Persaud D, Sarti E, Wilson M, Flisser A (1992) Neurocysticercosis in an Orthodox Jewish Community in New York City. N Engl J Med 327:692–695

Scholten T, Pang D, Lau TS (1976) Cysticercosis. Can Med Assoc J 115:612–613

Serpa JA, White AC Jr (2012) Neurocysticercosis in the United States. Pathog Global Health 106:256–260

Serpa JA, Graviss EA, Kass JS, White AC Jr (2011) Neurocysticercosis in Houston, Texas. An update. Medicine (Baltimore) 90:81–86

Sheth TN, Pilon L, Keystone J, Kucharczyk W (1998) Persistent MR contrast enhancement of calcified neurocysticercosis lesions. AJNR Am J Neuroradiol 19:79–82

Sorvillo FJ, Waterman SH, Richards FO, Schantz PM (1992) Cysticercosis surveillance: locally-acquired and travel-related infections and detection of intestinal tapeworm carriers in Los Angeles County. Am J Trop Med Hyg 47:365–371

Sorvillo FJ, DeGiorgio C, Waterman SH (2007) Deaths from cysticercosis, United States. Emerg Infect Dis 13:230–235

Sorvillo F, Wilkins P, Shafir S, Eberhard M (2011) Public health implications of cysticercosis acquired in the United States. Emerg Infect Dis 17:1–6

Townes JM, Hoffmann CJ, Kohn MA (2004) Neurocysticercosis in Oregon, 1995-2000. Emerg Infect Dis 10:508–510

Wadley JP, Shakir RA, Edwards JMR (2000) Experience with neurocysticercosis in the UK: correct diagnosis and neurosurgical management of the small enhancing brain lesion. Br J Neurosurg 14:211–218

Walker J, Chen S, Packham D, McIntyre P (1991) Five cases of neurocysticercosis diagnosed in Sydney. Southeast Asian J Trop Med Public Health 22(Suppl):242–244

Wallin MT, Kurtzke JF (2004) Neurocysticercosis in the United States: review of an important emerging infection. Neurology 63:1559–1564

White JC, Sweet WH, Richardson EP (1957) Cysticercosis cerebri. A diagnostic and therapeutic problem of increasing importance. N Engl J Med 256:479–486

Wickremesekera AC, Malham GM, Allen J, Macdonald GM (1996) Neurocysticercosis: a rare cause of epilepsy in New Zealand. N Z Med J 109:212–213

Yong JL, Warren BA (1994) Neurocysticercosis: a report of four cases. Pathology 26:244–249

Immunopathology of Taeniasis and Cysticercosis

6

The adaptive processes required for the parasite to survive inside the vertebrate host avoiding the attack of the immune system either as a tissue cyst or as an intestinal tapeworm, as well as the changes occurring after detection of the cysts by the immune system and the subsequent inflammatory reaction and parasite degeneration, result in a varied array of diverse mechanisms and scenarios which in most cases are only poorly understood.

6.1 Immunopathology of Human Cysticercosis

Our sources of knowledge for human cysticercosis are mostly clinical and necropsy cases, transversal studies in endemic populations (a few using neuroimaging), and occasional anecdotal data on epidemiological outbreaks or natural transmission scenarios (Dixon and Lipscomb 1961; Gajdusek 1978; Schantz et al. 1992). From these, we image human cysticercosis as a very common infection, in most cases resolving without apparent symptoms, which in a minority of cases presents with disabling clinical manifestations reflecting the specific characteristics of the involvement of the nervous system in each individual. In clinical settings, neurocysticercosis is by far more studied than any other location of cysts, and cysticercosis outside the nervous system is much less frequently seen. There are, however, many voids in our knowledge of the dynamics of transmission and the course of infection of human cysticercosis, including a very poor understanding of the role of the host's immune response (Fleury et al. 2010). Many variables affect the immune response; the most important ones seem to include the infective dose, previous exposures to the parasite, and the localization of the parasite.

In terms of the infective dose, we can only rely on cyst numbers as a proxy for the number of ingested oncospheres. This is obviously a distant surrogate since in most cases of human cysticercosis we only know the numbers of brain cysts which can be seen on brain imaging and do not know anything about cysts in the rest of the body (which should be several times more). Most ingested oncospheres will not result in established cysts. In pig infection models usually 2 % or less of the ingested

O.H. Del Brutto, H.H. García, *Cysticercosis of the Human Nervous System*,
DOI 10.1007/978-3-642-39022-7_6, © Springer-Verlag Berlin Heidelberg 2014

oncospheres result in visible cysts a few months later (Santamaria et al. 2002). The burden of infection will also affect the response. Clinical series demonstrate that individuals with a single brain lesion have much milder immune responses, while individuals with several cysts have more antigens and antibodies (García et al. 2010; Wilson et al. 1991).

We also know very few about how the innate or acquired immune responses will affect the likelihood of infection. It is very likely that oncospheres arriving into non-immunologically privileged sites be destroyed by the host immunity quite early (Flisser 1994). Moreover, some cysts may have established and resolved before clinical disease was evident. In endemic areas most individuals with neurocysticercosis (which may compose 15–20 % of the entire population) have only one or two calcified scars and no history of symptoms (Del Brutto et al. 2005; Fleury et al. 2003; Garcia-Noval et al. 1996; Montano et al. 2005; Sanchez et al. 1999). How an initial exposure will affect the likelihood of successful infection in further challenges is also not known.

In terms of localization, subarachnoid disease is associated to very high antigen levels and strong antibody responses (Fleury et al. 2011). Subarachnoid disease is characterized by marked extension of the parasite membrane with a hypertrophic response, and judging by the age profile of these patients, it is likely a late manifestation of neurocysticercosis (Bickerstaff et al. 1952). The strength of the immune response in parenchymal brain cysticercosis is directly related to the number of cysts and may be very weak or negative in patients with a single brain lesion (Wilson et al. 1991; Zini et al. 1990).

6.1.1 Single Enhancing Lesion

Travelers and children (Del Brutto 2013; Del Brutto et al. 2012) and native cases in the Indian subcontinent most frequently present with a single inflamed brain lesion, suggesting an early inflammatory response to a mild challenge (García et al. 2003b; Rajshekhar and Chandy 2000). In this view, a proportion of single lesion cases would correspond to cysticerci that resolved soon after infection, before full establishment and thus not provoking a strong immune response (García et al. 2010). The alternative hypothesis, that these result from existing viable cysts, collides with the very young age and mild antibody reactions which are typical in patients with single enhancing lesions.

6.1.2 Immune Response and Symptomatic Disease

Cysts get established soon after infection, likely in less than 3 months (Yoshino 1933). Most cysts in non-immunologically privileged sites are probably destroyed very fast by the host's immune response (see above under pig cysticercosis). Cysts in the brain and the eye survive preferentially due to the protection of the blood-brain barrier and the hemato-ocular barrier and manifest as symptomatic disease years after (Dixon and Lipscomb 1961).

In a series of detailed clinical descriptions of neurocysticercosis cases published in Germany and England in the early 1900s, several authors noted that subcutaneous parasites resolve years after than brain cysts and that neurological symptoms frequently occur between 3 and 5 years after infection (Dixon and Lipscomb 1961; Henneberg 1912; McArthur 1934). Live brain cysts can survive for many years without initiating the degeneration process. This long asymptomatic or oligosymptomatic stage likely results from active parasite mechanisms of immune evasion. Beyond preferential survival in immunoprotected cysts such as the brain or the eye, cysticercotic cysts may actually utilize molecular mimicry or mimetism mechanisms where the cyst covers itself with host immunoglobulins (Correa et al. 1985; Flisser 1989; Willms 2008), ingest and degrade host immunoglobulins (Damian 1987), inhibit the complement cascade at the C1q and C3b levels (Khan and Sotelo 1989), and secrete chemokines which modulate the immune attack of the host (Evans et al. 1998; Sciutto et al. 2007).

Frequently, symptoms and signs of neurocysticercosis appear or get exacerbated as a result of one or more cysts been attacked by the host immune system. This cellular attack consists of lymphocytes, plasma cells, and macrophages, frequently preceded by a well-defined eosinophil infiltrate, and gradually destroys the cyst (Flisser 1994; Mahanty and García 2010). In pathological specimens this process results in a colloidal stage of early destruction, in which the cysts contents become turbid and jelly, followed by a granular nodular stage in which no discernible liquid content is found anymore, to end as a calcified scar, as classically described by Escobar (Escobar 1983). In general, it is suggested that after an initial pro-inflammatory Th1 period with increased IL-12, TNF-a, and nitric oxide, the parasite establishes in a Th2-type response while asymptomatic, with low IFN-g and IgG2a antibodies and increased IgG1, IgE, IL-4, IL-13, and IL-15 (Alvarez et al. 2002; Bueno et al. 2004; Chavarria et al. 2003; Fleury et al. 2003; Restrepo et al. 2001; Rodriguez-Sosa et al. 2002; Terrazas et al. 1998; Villa and Kuhn 1996), possibly with alternatively activated macrophages as a contributory effector for this modulation (Gundra et al. 2011; Terrazas et al. 2005). It would eventually switch back to a Th1-like profile once the immune equilibrium is lost and the inflammatory response establishes and destroys the parasite, a process which is also triggered by treatment with antiparasitic agents. This inflammatory response causes most of the morbidity associated with neurocysticercosis (White et al. 1997) and responds very well to steroid treatment (Nash et al. 2006).

There is an apparent contrast between the benefit of inflammatory processes in eliminating the parasite and their deleterious effects in terms of clinical manifestations (Sciutto et al. 2007). This is particularly notorious when considering the effects of cyst death, natural of after antiparasitic treatment, in causing brain inflammation. Using steroids to suppress inflammation is required in both cases (Nash et al. 2011). Steroids have a very wide spectrum of action so its use does not contribute that much to deciphered specific inflammation mechanisms in neurocysticercosis. Long-term steroid treatment may lead to complications so judicious use is mandatory. Other generic anti-inflammatory therapies like methotrexate (Mitre et al. 2007) have been used to replace steroids, and immunosuppressive therapy has been used in severe inflammatory cases (Del Brutto and Sotelo 1988). Whether more specific agents such as anti-TNF may be of use in neurocysticercosis remains to be explored.

6.1.3 Diagnostic Antigens and Immunodiagnosis

A large series of specific antigens and epitopes have been identified for *Taenia solium* (Deckers and Dorny 2010; Rodriguez et al. 2012). Some of them are associated with protection to infection and thus have been used to develop pig vaccines, and most others were identified as potential diagnostic antigens.

6.1.4 Immunity to Cysticercosis and Comorbidity

Some authors have reported an increased presence of immune diseases with alteration of the immune function in patients with neurocysticercosis (Sanz 1987), and an association between neurocysticercosis and cerebral glioma has been reported both as isolated cases (Agapejev et al. 1992) and in a controlled study (Del Brutto et al. 1997). The underlying hypothesis is that inflammation and suppression of local immune response may result in malignant transformation of glial cells, although the association has not been further studied.

6.1.5 NCC in Immunosuppressed Patients

A few cases of neurocysticercosis in transplant recipients have been reported (Barra Valencia et al. 2007; Gordillo-Paniagua et al. 1987; Hoare et al. 2006). Neurocysticercosis is rarely seen in series of patients with HIV infection, most likely as a coincidental association (Moskowitz et al. 1984; Soto Hernandez et al. 1996; Thornton et al. 1992; White et al. 1995). Some authors suggest that extraparenchymal locations including giant cysts would be more common in HIV-infected individuals, but this has not been confirmed in larger series (Soto Hernandez et al. 1996; Walker and Zunt 2005).

6.2 Immunopathology of *Taenia solium* Taeniasis

Relatively little information exist on the human infection with the adult intestinal tapeworm and the host's immunological response.

6.2.1 Tapeworm Prevalence and Life Span

Most of the literature is composed of prevalence studies of taeniasis concluding that the population prevalence of intestinal taeniasis in endemic countries locates around 1 %, usually ranging from 0.5 to 2 % (Allan et al. 1996; García et al. 2003a; Sarti et al. 1992, 1994). Most of these surveys determined only the prevalence of adult tapeworms detected by stool microscopy and are likely underestimations of the prevalence because of the suboptimal sensitivity of microscopy and also because they will miss early infections. Parasite antigens can be detected in stools since early in the infection, 3 weeks, and studies using coproantigen detection find between 1.5 and 3 times more tapeworms (Allan et al. 1990; García et al. 2003a). The discordance between coproantigen and microscopy data suggests that either some or many infections establish but resolve before egg excretion, mature tapeworms may have periods where no eggs are excreted, or both of the above.

The tapeworm life span of *T. solium* was assumed to be of more than 20 years, based on particular cases of long-standing *Taenia saginata* infection (Faust et al. 1984). However, clinical and epidemiological data including age prevalence curves suggests that the adult tapeworm lives less than 5 years (Allan et al. 1996).

6.2.2 Tapeworm Infection

The worm establishes in the small intestine by anchoring its scolex in the mucosa using its four suckers and its double crown of hooks, with a local inflammatory response in the area which is composed by macrophages, plasma cells, lymphocytes, fibroblasts, neutrophils, and eosinophils, with local histamine release (Avila et al. 2002; Willms 2008). The local response may reduce the likelihood of tapeworm establishment and thus it has been explored as a possible vaccine (Leon-Cabrera et al. 2009, 2012). So far partial protection in a rodent model has been obtained. Other vaccines for intestinal cestodes are also being developed, including a dog vaccine for *Echinococcus granulosus* (Zhang et al. 2006).

Two specific diagnostic antigens specific to the adult tapeworm stage have been identified, namely, T33 and T38 (Levine et al. 2004; Verastegui et al. 2003). These antibodies are present in the sera of 95 % or more tapeworm carriers. However, the antibody response seems to outlast the intestinal tapeworm for a long period, hampering its use as a diagnostic test.

There is little evidence on whether reinfection with a new adult *Taenia solium* is possible. This is quite difficult to determine since it requires certainty that the patient was cured (meaning an accurate follow-up with negative testing for 3 months or more). From a large series of around 300 patients with *T. solium* taeniasis and also confirmed cure, only three cases of reinfection were seen by our group in a median of 2 years of follow-up (Cysticercosis Working Group in Peru, 2013, unpublished data).

6.3 Immunopathology of Porcine Cysticercosis

Pigs live much less than humans and are likely exposed to much heavier doses of infective eggs. In field conditions, a minority of pigs ingest eggs directly from the stools of a tapeworm carrier who defecates in the open field, resulting in heavy infections. Most animals, however, harbor only a few cysts in the entire carcass (Huerta et al. 2002). Dispersion mechanisms have been postulated to explain this apparent discrepancy including pig-to-pig transmission and dung beetles as egg carriers (Gonzalez et al. 2006; Juris et al. 1995; Lonc 1980).

The parasite seems to be well adapted to the porcine host and apparently causes minimal if any symptomatic disease (de Aluja and Villalobos 2000). Histopathological studies in the porcine host show a spectrum of severity of the perilesional inflammatory reaction which ranges from almost no discernible response to severe inflammation with complete parasitic degeneration (Londono et al. 2002).

Antibody responses in pigs have been assessed using the specific EITB assay and show that sizable numbers of pigs (frequently more than half of all pigs in an endemic community) may have specific antibodies (García et al. 2003a). A proportion of these circulating antibody responses correspond to persisting maternal antibodies passively transferred on the sow's milk (Gonzalez et al. 1999). Strong responses are less frequent. Circulating antigen can be measured by antigen capture ELISA using monoclonal antibodies (Ganaba et al. 2011). However, antigen levels in feral pigs are more difficult to interpret because of cross-reactive responses with *Taenia hydatigena*, a common tapeworm infection.

The interpretation of porcine serological data in comparison to human data should always take into account that the information available for human infections is mostly restricted to brain cysts only, while in pigs the complete parasite burden is assessed (in the entire carcass). Brain cysts should be a minority of all cysts in the individual and likely those surviving because of the blood-brain barrier protection (de Aluja and Villalobos 2000). Pigs with only degenerated cysts resulting from the host immunity overcoming the infection are also a common finding despite the fact that most pigs are sacrificed at 9 months or before. In general, the proportion of pigs with viable cysts is several times less than the proportion of antibody-positive pigs.

Vaccines against porcine cysticercosis have been developed and successfully tested in controlled and field conditions, particularly using the TSOL18 antigen (Flisser et al. 2004; Gonzalez et al. 2005). This is an oncospheral antigen which confers over 99 % protection when given in two intramuscular doses. Other vaccines used a heterologous *T. crassiceps* antigen (Huerta et al. 2002) or DNA immunization (Guo et al. 2007). Porcine immunization can still be optimized by reducing the administration to a single dose or developing an oral vaccine.

References

Agapejev S, Alves A, Zanini MA, Ueda AK, Pereira EM (1992) Oligodendroglioma cistico e positividade das reaçoes para cisticercose. Arq Neuropsiquiatr 50:234–238

Allan JC, Avila G, Garcia Noval J, Flisser A, Craig PS (1990) Immunodiagnosis of taeniasis by coproantigen detection. Parasitology 101(Pt 3):473–477

Allan JC, Velasquez-Tohom M, Garcia-Noval J et al (1996) Epidemiology of intestinal taeniasis in four, rural, Guatemalan communities. Ann Trop Med Parasitol 90(2):157–165

Alvarez JI, Colegial CH, Castano CA, Trujillo J, Teale JM, Restrepo BI (2002) The human nervous tissue in proximity to granulomatous lesions induced by *Taenia solium* metacestodes displays an active response. J Neuroimmunol 127(1–2):139–144

Avila G, Aguilar L, Benitez S, Yepez-Mulia L, Lavenat I, Flisser A (2002) Inflammatory responses in the intestinal mucosa of gerbils and hamsters experimentally infected with the adult stage of *Taenia solium*. Int J Parasitol 32(10):1301–1308

Barra Valencia V, Moreno Elola-Olaso A, Fundora Suarez Y et al (2007) Second case of neurocysticercosis in a patient with liver transplantation (first case in Spain): a case report. Transplant Proc 39(7):2454–2457

Bickerstaff ER, Cloake PCP, Hughes B, Smith WT (1952) The racemose form of cerebral cysticercosis. Brain 75:1–16

Bueno EC, dos Ramos Machado L, Livramento JA, Vaz AJ (2004) Cellular immune response of patients with neurocysticercosis (inflammatory and non-inflammatory phases). Acta Trop 91(2):205–213

Chavarria A, Roger B, Fragoso G et al (2003) TH2 profile in asymptomatic *Taenia solium* human neurocysticercosis. Microbes Infect 5(12):1109–1115

Correa D, Dalma D, Espinoza B et al (1985) Heterogeneity of humoral immune components in human cysticercosis. J Parasitol 71(5):535–541

Damian RT (1987) The exploitation of host immune responses by parasites. J Parasitol 73(1):3–13

de Aluja AS, Villalobos MN (2000) *Taenia solium* cysticercosis in pigs in Mexico. Vet Mex 31(3):239–244

Deckers N, Dorny P (2010) Immunodiagnosis of *Taenia solium* taeniosis/cysticercosis. Trends Parasitol 26(3):137–144

Del Brutto OH (2013) Neurocysticercosis in infants and toddlers: report of seven cases and review of published patients. Pediatr Neurol 48(6):432–435. doi:10.1016/j.pediatrneurol.2013.02.001

Del Brutto OH, Castillo PR, Mena IX, Freire AX (1997) Neurocysticercosis among patients with cerebral gliomas. Arch Neurol 54:1125–1128

Del Brutto OH, Sotelo J (1988) Neurocysticercosis: an update. Rev Infect Dis 10(6):1075–1087

Del Brutto OH, Santibañez R, Idrovo L et al (2005) Epilepsy and neurocysticercosis in Atahualpa: a door-to-door survey in rural coastal Ecuador. Epilepsia 46(4):583–587

Del Brutto OH, Nash TE, García HH (2012) Cysticerci-related single parenchymal brain enhancing lesions in non-endemic countries. J Neurol Sci 319(1–2):32–36. doi:10.1016/j.jns.2012.05.027

Dixon HB, Lipscomb FM (1961) Cysticercosis: an analysis and follow-up of 450 cases, vol 299. Medical Research Council, London

Escobar A (1983) The pathology of neurocysticercosis. In: Palacios E, Rodriguez-Carbajal J, Taveras JM (eds) Cysticercosis of the central nervous system. Charles C. Thomas, Springfield, pp 27–54

Evans CA, García HH, Hartnell A et al (1998) Elevated concentrations of eotaxin and interleukin-5 in human neurocysticercosis. Infect Immun 66(9):4522–4525

Faust EC, Russell PF, Jung RC (1984) Parasitología clínica. Salvat, Barcelona

Fleury A, Gomez T, Alvarez I et al (2003) High prevalence of calcified silent neurocysticercosis in a rural village of Mexico. Neuroepidemiology 22(2):139–145

Fleury A, Escobar A, Fragoso G, Sciutto E, Larralde C (2010) Clinical heterogeneity of human neurocysticercosis results from complex interactions among parasite, host and environmental factors. Trans R Soc Trop Med Hyg 104(4):243–250

Fleury A, Carrillo-Mezo R, Flisser A, Sciutto E, Corona T (2011) Subarachnoid basal neurocysticercosis: a focus on the most severe form of the disease. Expert Rev Anti Infect Ther 9(1):123–133

Flisser A (1989) *Taenia solium* cysticercosis: some mechanisms of parasite survival in immunocompetent hosts. Acta Leiden 57(2):259–263

Flisser A (1994) Taeniasis and cysticercosis due to *Taenia solium*. Prog Clin Parasitol 4:77–116

Flisser A, Gauci CG, Zoli A et al (2004) Induction of protection against porcine cysticercosis by vaccination with recombinant oncosphere antigens. Infect Immun 72(9):5292–5297

Gajdusek DC (1978) Introduction of *Taenia solium* into west New Guinea with a note on an epidemic of burns from cysticercus epilepsy in the Ekari people of the Wissel Lakes area. P N G Med J 21(4):329–342

Ganaba R, Praet N, Carabin H et al (2011) Factors associated with the prevalence of circulating antigens to porcine cysticercosis in three villages of Burkina Faso. PLoS Negl Trop Dis 5(1):e927

García HH, Gilman RH, Gonzalez AE et al (2003a) Hyperendemic human and porcine *Taenia solium* infection in Peru. Am J Trop Med Hyg 68(3):268–275

García HH, Gonzalez AE, Gilman RH, for The Cysticercosis Working Group in Peru (2003b) Diagnosis, treatment and control of *Taenia solium* cysticercosis. Curr Opin Infect Dis 16:411–419

García HH, Gonzalez AE, Rodriguez S et al (2010) Neurocysticercosis: unraveling the nature of the single cysticercal granuloma. Neurology 75(7):654–658

Garcia-Noval J, Allan JC, Fletes C et al (1996) Epidemiology of *Taenia solium* taeniasis and cysticercosis in two rural Guatemalan communities. Am J Trop Med Hyg 55(3):282–289

Gonzalez AE, Verastegui M, Noh JC et al (1999) Persistence of passively transferred antibodies in porcine *Taenia solium* cysticercosis. Cysticercosis Working Group in Peru. Vet Parasitol 86(2):113–118

Gonzalez AE, Gauci CG, Barber D et al (2005) Vaccination of pigs to control human neurocysticercosis. Am J Trop Med Hyg 72(6):837–839

Gonzalez AE, López-Urbina T, Tsang B et al (2006) Transmission dynamics of *Taenia solium* and potential for pig-to-pig transmission. Parasitol Int 55(Suppl):S131–S135

Gordillo-Paniagua G, Munoz-Arizpe R, Ponsa-Molina R (1987) Unusual complication in a patient with renal transplantation: cerebral cysticercosis. Nephron 45(1):65–67

Gundra UM, Mishra BB, Wong K, Teale JM (2011) Increased disease severity of parasite-infected TLR2-/- mice is correlated with decreased central nervous system inflammation and reduced numbers of cells with alternatively activated macrophage phenotypes in a murine model of neurocysticercosis. Infect Immun 79(7):2586–2596

Guo A, Jin Z, Zheng Y et al (2007) Induction of protection against porcine cysticercosis in growing pigs by DNA vaccination. Vaccine 25(1):170–175

Henneberg R (1912) Die tierischen parasiten des zentralnervensystem. In: Lewandowsky M (ed) Handbuch der neurologie, vol III. Verlag Von Julius Springer, Berlin, pp 643–712

Hoare M, Gelson WT, Antoun N, Alexander GJ (2006) Early recurrence of neurocysticercosis after orthotropic liver transplant. Liver Transpl 12(3):490–491

Huerta M, de Aluja AS, Fragoso G et al (2002) Synthetic peptide vaccine against *Taenia solium* pig cysticercosis: successful vaccination in a controlled field trial in rural Mexico. Vaccine 20(1–2):262–266

Juris P, Vilagiova I, Plachy P (1995) The importance of flies (Diptera-Brachycera) in the dissemination of helminth eggs from sewage treatment plants. Vet Med (Praha) 40(9):289–292

Khan NA, Sotelo J (1989) Presentation of a membrane cysticercus antigen and its homology with excretory – secretory antigen. Acta Leiden 57(2):123–129

Leon-Cabrera S, Cruz-Rivera M, Mendlovic F et al (2009) Standardization of an experimental model of human taeniosis for oral vaccination. Methods 49(4):346–350

Leon-Cabrera S, Cruz-Rivera M, Mendlovic F et al (2012) Immunological mechanisms involved in the protection against intestinal taeniosis elicited by oral immunization with *Taenia solium* calreticulin. Exp Parasitol 132(3):334–340

Levine MZ, Calderon JC, Wilkins PP et al (2004) Characterization, cloning, and expression of two diagnostic antigens for *Taenia solium* tapeworm infection. J Parasitol 90(3): 631–638

Lonc E (1980) The possible role of the soil fauna in the epizootiology of cysticercosis in cattle. I. Earthworms – the biotic factor in a transmission of Taenia saginata eggs. Angew Parasitol 21(3):133–139

Londono DP, Alvarez JI, Trujillo J, Jaramillo MM, Restrepo BI (2002) The inflammatory cell infiltrates in porcine cysticercosis: immunohistochemical analysis during various stages of infection. Vet Parasitol 109(3–4):249–259

Mahanty S, García HH (2010) Cysticercosis and neurocysticercosis as pathogens affecting the nervous system. Prog Neurobiol 91(2):172–184

McArthur WP (1934) Cysticercosis as seen in the British army with special reference to the production of epilepsy. Trans R Soc Trop Med Hyg 27:343–363

Mitre E, Talaat KR, Sperling MR, Nash TE (2007) Methotrexate as a corticosteroid-sparing agent in complicated neurocysticercosis. Clin Infect Dis 44(4):549–553

Montano SM, Villaran MV, Ylquimiche L et al (2005) Neurocysticercosis: association between seizures, serology, and brain CT in rural Peru. Neurology 65(2):229–233

Moskowitz LB, Hensley GT, Chan JC, Gregorios J, Conley FK (1984) The neuropathology of acquired immune deficiency syndrome. Arch Pathol Lab Med 108(11):867–872

Nash TE, Singh G, White AC et al (2006) Treatment of neurocysticercosis: current status and future research needs. Neurology 67(7):1120–1127

Nash TE, Mahanty S, García HH (2011) Corticosteroid use in neurocysticercosis. Expert Rev Neurother 11(8):1175–1183

Rajshekhar V, Chandy MJ (2000) Solitary cysticercus granuloma: the disappearing lesion. Orient Longman, Chennai

Restrepo BI, Aguilar MI, Melby PC, Teale JM (2001) Analysis of the peripheral immune response in patients with neurocysticercosis: evidence for T cell reactivity to parasite glycoprotein and vesicular fluid antigens. Am J Trop Med Hyg 65(4):366–370

Rodriguez S, Wilkins P, Dorny P (2012) Immunological and molecular diagnosis of cysticercosis. Pathog Glob Health 106(5):286–298

Rodriguez-Sosa M, David JR, Bojalil R, Satoskar AR, Terrazas LI (2002) Cutting edge: susceptibility to the larval stage of the helminth parasite *Taenia crassiceps* is mediated by Th2 response induced via STAT6 signaling. J Immunol 168(7):3135–3139

Sanchez AL, Lindback J, Schantz PM et al (1999) A population-based, case-control study of *Taenia solium* taeniasis and cysticercosis. Ann Trop Med Parasitol 93(3):247–258

Santamaria E, Plancarte A, de Aluja AS (2002) The experimental infection of pigs with different numbers of *Taenia solium* eggs: immune response and efficiency of establishment. J Parasitol 88(1):69–73

Sanz CR (1987) Host response in childhood neurocysticercosis. Some pathological aspects. Childs Nerv Syst 3(4):206–207

Sarti E, Schantz PM, Plancarte A et al (1992) Prevalence and risk factors for *Taenia solium* taeniasis and cysticercosis in humans and pigs in a village in Morelos, Mexico. Am J Trop Med Hyg 46(6):677–685

Sarti E, Schantz PM, Plancarte A et al (1994) Epidemiological investigation of *Taenia solium* taeniasis and cysticercosis in a rural village of Michoacan state, Mexico. Trans R Soc Trop Med Hyg 88(1):49–52

Schantz PM, Moore AC, Munoz JL et al (1992) Neurocysticercosis in an Orthodox Jewish community in New York City. N Engl J Med 327(10):692–695

Sciutto E, Chavarria A, Fragoso G, Fleury A, Larralde C (2007) The immune response in *Taenia solium* cysticercosis: protection and injury. Parasite Immunol 29(12):621–636

Soto Hernandez JL, Ostrosky Zeichner L, Tavera G, Gomez Avina A (1996) Neurocysticercosis and HIV infection: report of two cases and review. Surg Neurol 45(1):57–61

Terrazas LI, Bojalil R, Govezensky T, Larralde C (1998) Shift from an early protective Th1-type immune response to a late permissive Th2-type response in murine cysticercosis (Taenia crassiceps). J Parasitol 84(1):74–81

Terrazas LI, Montero D, Terrazas CA, Reyes JL, Rodriguez-Sosa M (2005) Role of the programmed Death-1 pathway in the suppressive activity of alternatively activated macrophages in experimental cysticercosis. Int J Parasitol 35(13):1349–1358

Thornton CA, Houston S, Latif AS (1992) Neurocysticercosis and human immunodeficiency virus infection. A possible association. Arch Neurol 49(9):963–965

Verastegui M, Gilman RH, García HH et al (2003) Prevalence of antibodies to unique *Taenia solium* oncosphere antigens in taeniasis and human and porcine cysticercosis. Am J Trop Med Hyg 69(4):438–444

Villa OF, Kuhn RE (1996) Mice infected with the larvae of Taenia crassiceps exhibit a Th2-like immune response with concomitant anergy and downregulation of Th1- associated phenomena. Parasitology 112(Pt 6):561–570

Walker M, Zunt JR (2005) Parasitic central nervous system infections in immunocompromised hosts. Clin Infect Dis 40(7):1005–1015

White AC Jr, Dakik H, Diaz P (1995) Asymptomatic neurocysticercosis in a patient with AIDS and cryptococcal meningitis. Am J Med 99(1):101–102

White AC Jr, Robinson P, Kuhn R (1997) *Taenia solium* cysticercosis: host-parasite interactions and the immune response. Chem Immunol 66:209–230

Willms K (2008) Morphology and biochemistry of the pork tapeworm, *Taenia solium*. Curr Top Med Chem 8(5):375–382

Wilson M, Bryan RT, Fried JA et al (1991) Clinical evaluation of the cysticercosis enzyme-linked immunoelectrotransfer blot in patients with neurocysticercosis. J Infect Dis 164(5): 1007–1009

Yoshino K (1933) Studies on the post-embryonal development of *Taenia solium*: III. On the development of Cysticercus cellulosae within the definitive intermediate host. J Med Assoc Formosan 32:166–169

Zhang W, Zhang Z, Shi B et al (2006) Vaccination of dogs against Echinococcus granulosus, the cause of cystic hydatid disease in humans. J Infect Dis 194(7):966–974

Zini D, Farrell VJ, Wadee AA (1990) The relationship of antibody levels to the clinical spectrum of human neurocysticercosis. J Neurol Neurosurg Psychiatry 53(8):656–661

Clinical Manifestations of Neurocysticercosis

7

Defining a typical clinical syndrome of neurocysticercosis is not possible. The disease may be asymptomatic or may present with a number of nonspecific manifestations, such as seizures, headaches, focal neurological deficits, increased intracranial pressure, or cognitive decline (Carabin et al. 2011; Del Brutto 2012; Sotelo 2011). This pleomorphism is mainly related to individual differences in the number and location of the lesions within the nervous system, as well as to the severity of the host's immune response against the parasite (Fleury et al. 2010).

Cysticercosis affects indistinctly people from birth to senility. However, mean age of patients in most large series has been between 25 and 40 years (Croker et al. 2012; Sotelo et al. 1985). Until recently, neurocysticercosis was considered rare in infants and children. However, the introduction of modern imaging diagnostic methods has resulted in increased recognition of pediatric cases. Most important, the course of the disease is different in infants and children compared to adults (Basu et al. 2007; Del Brutto 2013; Talukdar et al. 2002). Regarding gender, most studies have shown that neurocysticercosis equally affects men and women. Nevertheless, it has been suggested that the disease tends to be more severe in women. It is possible that the increased reactivity of the immune system in women could be partially responsible for the observed gender differences in the severity of neurocysticercosis (Del Brutto et al. 1988a; Fleury et al. 2004).

Geographical differences in the clinical spectrum of the disease have also been noted. Indeed, disseminated infections associated with symptomatic subcutaneous and muscular cysticercosis are observed more frequently in Africa and Asia than in the Americas (Bhalla et al. 2008). Although no antigenic differences between cysticerci obtained from Asiatic and American patients have been demonstrated, the higher prevalence of symptomatic disseminated cysticercosis among Asians and Africans has led to the suggestion of possible strain differences (Craig et al. 1996). A more plausible explanation to this finding is that more severe burden of infections due to environmental, cultural, and dietary differences, as well as genetic variations between populations, may explain clinical differences between African, Asian, and American patients with neurocysticercosis.

O.H. Del Brutto, H.H. García, *Cysticercosis of the Human Nervous System*,
DOI 10.1007/978-3-642-39022-7_7, © Springer-Verlag Berlin Heidelberg 2014

7.1 Parenchymal Neurocysticercosis

7.1.1 Seizures

Seizures are the most common clinical manifestation of parenchymal brain cysticercosis and usually represent the primary or sole manifestation of this form of the disease (Del Brutto et al. 1992). While seizure prevalence among patients with neurocysticercosis varies according to the population studied and the diagnostic methods used, most series have shown that seizures occur in up to 80 % of cases (Ndimubanzi et al. 2010).

Epileptogenesis in neurocysticercosis has been a subject of debate. It was initially believed that neurocysticercosis-related seizures occurred almost exclusively when parasites began to degenerate and that the pathological substract explaining the occurrence of the seizure disorder was the acute inflammation surrounding dying cysticerci. Further studies, however, showed that seizures may occur at any stage of cysticerci involution within the brain parenchyma, from viable cysts to calcifications. In the latter, the intense gliosis that develops around dying or already dead cysticerci, as well as periodic morphological remodeling of calcification with exposure of antigenic material to the brain parenchyma, or even the development of hippocampal sclerosis may be the factors accounting for the epileptogenic activity of calcified cysticerci (Gupta et al. 2012; Nash et al. 2004; Rathore and Radhakrishnan 2012; Rathore et al. 2012).

Neurocysticercosis is a major cause of acquired epilepsy in most of the developing world and has been considered as the single cause explaining the increased incidence and prevalence of epilepsy in these regions (Blocher et al. 2011; Carabin et al. 2011; Del Brutto et al. 2005; Medina et al. 2005; Preux and Druet-Cabanac 2005; Villarán et al. 2009). In addition, this parasitic disease has also been considered one of the leading cause of adult-onset epilepsy (seizures starting in individuals aged 25 years or more) in areas of the world where the disease is endemic, particularly in developing countries of Latin America, Asia, and Africa. By the end of the twentieth century, large series of patients with adult-onset epilepsy from Brazil (Arruda 1991), Ecuador (Del Brutto and Noboa 1991), Mexico (Medina et al. 1990), India (Chopra et al. 1981), and South Africa (van As and Joubert 1991) demonstrated that neurocysticercosis occurred in 25–50 % of the cases, representing the single most common cause of this syndrome. On the other hand, there is more recent evidence suggesting that neurocysticercosis may no longer be the most common cause of symptomatic adult-onset epilepsy in patients evaluated in, at least, some urban centers of disease-endemic countries (Del Brutto and Del Brutto 2012a; Suástegui et al. 2009). In such regions, neurocysticercosis is now outnumbered by cerebrovascular diseases and brain tumors which appear to be the leading causes of acquired epilepsy in adult populations. Two factors may be responsible for the reported drop in the number of patients with neurocysticercosis-related adult-onset epilepsy, including increased sanitation and widespread availability of neuroimaging machines which allow early recognition of neurocysticercosis cases by general practitioners who can start prompt cysticidal drug therapy, avoiding long-term sequelae of chronically untreated disease such as poor seizure control.

Seizures due to neurocysticercosis are most often generalized tonic-clonic or simple partial with motor symptomatology, although some patients may present with complex partial seizures, myoclonic seizures, truncal seizures, or even with specific epileptic syndromes such as the Landau-Kleffner syndrome. The type of seizures has been considered to be related to the number and topographic location of the parasites, whereby patients with a single lesion present with partial seizures, while patients with multiple lesions have tonic-clonic generalized seizures (Sotelo et al. 1985). However, a study of 203 patients with epilepsy due to neurocysticercosis showed no difference in the frequency of partial seizures in patients with single parenchymal brain cysts as compared with those with multiple cysts (Del Brutto et al. 1992).

7.1.2 Focal Neurological Deficits

From 5 to 15 % of patients with parenchymal brain cysticercosis present focal signs in the neurological examination (Carabin et al. 2011). While weakness associated with pyramidal tract signs predominate, almost every focal neurological sign has been described in these patients, including sensory deficits, language disturbances, involuntary movements (hemichorea, tremor, facial myokymia, and blepharospasm), parkinsonian rigidity, gait disturbances, and signs of brainstem dysfunction (Keane 1982; McCormick 1985; Sotelo et al. 1985). As expected, focal signs are related to parasites strategically located in eloquent areas of the brain parenchyma, particularly in the cerebral cortex or the brainstem (Del Brutto and Del Brutto 2013a). Interestingly, parasites located in the basal ganglia are most often clinically silent (Cosentino et al. 2002). Focal neurological deficits in patients with parenchymal brain cysticercosis usually follow a subacute or chronic course resembling that of a brain tumor or focal granulomatous process such as an intracranial tuberculoma. However, some patients present with a stroke-like onset and rapid development of focal signs resembling a cerebrovascular event (Del Brutto 2012).

7.1.3 Increased Intracranial Pressure

Different pathogenetic mechanisms explain the occurrence of increased intracranial pressure in patients with parenchymal neurocysticercosis, the most common being the so-called cysticercotic encephalitis (Rangel et al. 1987). This severe form of the disease most often affects children and young women and results from the acute inflammatory response that ensues in some patients as the result of massive cysticercal infection of the brain parenchyma. Intracranial hypertension in these cases is related to diffuse brain edema and not due to the physical presence of the parasites, since massive cysticercal infection of the brain parenchyma is not associated with increased intracranial pressure in the absence of immune reaction against the parasites (Del Brutto and Campos 2012). Patients with cysticercotic encephalitis present with a syndrome of acute or subacute onset characterized by clouding of consciousness, seizures, decreased visual acuity, headache, vomiting, and papilledema. These patients are severely ill during the acute phase of the disease (some patients may die as the result of intracranial hypertension), but those who survive the acute episode usually recover without sequelae (Noboa 1992).

Other mechanisms explaining the development of intracranial hypertension in patients with parenchymal brain cysticercosis include the rare presence of a large parenchymal brain cyst that displaces midline structures or the obstruction of CSF circulation due to a midbrain cyst that obliterates the cerebral aqueduct (Poon et al. 1980).

7.1.4 Cognitive Decline and Psychiatric Disturbances

That patients with neurocysticercosis may develop psychiatric manifestations or organic mental disorders has long been known. Of historical importance, the first case of neurocysticercosis reported from Asia came from a man who had lived in a lunatic asylum in Madras (Wadia 1995), and it was not uncommon in Mexico City to find patients who, after being admitted to psychiatric hospitals for years, had a correct diagnosis of neurocysticercosis when they developed seizures or focal neurological signs (Nieto 1956).

The spectrum of cognitive impairment in patients with neurocysticercosis is wide and their prevalence has varied from one series to another. Recent studies have shown that a sizable proportion of these patients have a decrease in quality of life, depression, or mild cognitive impairment that is not directly related to the severity or the specific form of the disease (Bhattarai et al. 2011; de Andrade et al. 2010; Monteiro de Almeida and Gurjão 2011; Wallin et al. 2012). This contrasts with previous studies suggesting that these manifestations correlate with the presence of increased intracranial pressure but not with parenchymal brain cysts or with the coexistence of seizures (Forlenza et al. 1997). Overt dementia has been found in 6–15 % of neurocysticercosis patients, a finding that seems to be related with the presence of active disease (Ramirez-Bermudez et al. 2005; Rodrigues et al. 2012).

Psychotic episodes characterized by confusion, paranoid ideation, psychomotor agitation, violent behavior, and visual hallucinations have also been described in patients with parenchymal neurocysticercosis (Shandera et al. 1994; Shriqui and Milette 1992). Some of these episodes could represent attacks of psychomotor epilepsy or postictal psychosis (Feinstein et al. 1990).

7.1.5 Headache

The occurrence of headache without any other evidence of increased intracranial pressure has been reported in several series of neurocysticercosis patients with a prevalence that may be as high as 30 % of cases (Carabin et al. 2011). Such headaches are intermittent and resemble those seen in migraine, tension-type headache, or other primary headache disorders, raising the question of whether they are related to cysticercosis or may occur as an incidental finding in areas where both conditions are highly prevalent (Finsterer et al. 2006; Mishra 2007; Rajshekhar 2000). An old case-control study suggested that the prevalence of migraine-type headaches was significantly increased among patients with neurocysticercosis (Cruz et al. 1995). However, until recently, there was no convincing evidence of a cause-and-effect relationship between these two conditions. A large, case-control study comparing the prevalence of neurocysticercosis among patients with major neurological disorders found a significantly increased prevalence of parenchymal brain calcifications among patients with an otherwise typical "primary headache disorder" when compared with that seen in other neurological disorders (Del Brutto and Del Brutto 2012b). It is possible that periodic remodeling of intracranial calcifications exposing parasitic antigenic material to the host causes transient inflammatory changes in the brain parenchyma that may be the cause of transient and recurrent headache episodes mimicking a primary headache disorder, thus providing a rationale for the association between headache and calcified neurocysticercosis.

7.2 Subarachnoid Neurocysticercosis

7.2.1 Increased Intracranial Pressure

Communicating hydrocephalus is the most common pathogenetic mechanism explaining the occurrence of increased intracranial pressure in patients with subarachnoid neurocysticercosis. Hydrocephalus may occur by extension of the subarachnoid inflammatory reaction to the leptomeninges at the base of the brain occluding the foramina of Luschka and Magendie or by chronic inflammation and fibrosis of the arachnoid villi causing obstruction to the reabsorption of CSF (Lobato et al. 1981). In these patients, the syndrome of increased intracranial pressure without focalizing signs presents with a subacute onset and progressive course, although some patients may follow a chronic and normotensive course (Revuelta-Gutierrez and Valdéz-García 1995). Hydrocephalus is an ominous sign associated with a 50 % mortality rate at 2 years (Sotelo and Marin 1987).

Development of large cysts at the Sylvian fissure or CSF cisterns at the base of the brain (particularly the cerebellopontine angle cistern) are often associated with focal neurological deficits and signs and symptoms of increased intracranial pressure of subacute onset and progressive course that resemble the clinical manifestations of intracranial tumors or other focal granulomatous infections of the central nervous system (Fleury et al. 2011; Kelesidis and Tsiodras 2012).

7.2.2 Cranial Nerve Involvement

Cysticercotic arachnoiditis causes entrapment of cranial nerves arising from the ventral aspect of the brainstem. The oculomotor nerves, running a long course within the basal subarachnoid space from their origin until their entrance into the cavernous sinuses, are often enclosed within the dense inflammatory exudate that occurs in this form of the disease. Extraocular muscle paralysis, diplopia, and pupillary abnormalities are the result of this entrapment (Keane 1993). The optic nerves and the optic chiasm may also be encased within this exudate with the subsequent development of diminution of the visual acuity or visual field defects (Keane 1982).

7.2.3 Acute Meningitis

The absence of fever and signs of meningeal irritation have been considered useful for the differential diagnosis between cysticercosis and other causes of meningitis such as that caused by pyogenic microorganisms, *Mycobacterium tuberculosis* or fungi. However, the clinical course of some patients with cysticercotic meningitis is acute and may be associated with fever and neck stiffness. So, this parasitic disease should also be included in the differential diagnosis of acute meningitis in endemic regions (Bonametti et al. 1994; Srinivas and Chandramukhi 1992).

7.2.4 Stroke Syndromes

A stroke may occur in up to 12 % of patients with neurocysticercosis, particularly in those with the subarachnoid form of the disease (Marquez and Arauz 2012). It must be noted, however, that in endemic regions, a patient with neurocysticercosis may have a stroke from unrelated reasons. Therefore, the diagnosis of neurocysticercosis-related stroke is only possible after proper interpretation of data provided by neuroimaging studies, results of immunologic tests, CSF findings, and the evaluation of other risk factors for cerebrovascular disease that may be present in a given patient (Del Brutto 1992).

A variety of stroke syndromes, including transient ischemic attacks, cerebral infarctions, and intracranial hemorrhages, have been described in patients with neurocysticercosis (Table 7.1). Different stroke subtypes are related to different pathogenetic mechanisms and produce varied clinical manifestations (Cantú and Barinagarrementeria 1996). Transient ischemic attacks have been occasionally reported and are most often related to intermittent stenosis of major intracranial arteries secondary to meningeal cysticerci engulfing such vessels. A sizable proportion of these patients eventually develop a cerebral infarction when the inflammatory process occludes the affected artery (Aditya et al. 2004; McCormick et al. 1983).

Cerebral infarctions occur as the result of inflammatory occlusion of small penetrating branches of major intracranial arteries or of occlusion of major vessels. Small infarctions may be located in the subcortical white matter, the internal capsule, or the brainstem and present as lacunar syndromes (ataxic hemiparesis, pure motor hemiparesis, sensorimotor stroke) indistinguishable from those caused by atherosclerosis (Barinagarrementeria and Del Brutto 1989). Large cerebral infarctions, related to the occlusion of the anterior or middle cerebral artery, cause focal neurological signs that may be associated with cognitive decline when both anterior cerebral arteries are affected (Monteiro et al. 1994; Rodriguez-Carbajal et al. 1989).

Hemorrhagic strokes have been reported in a dozen of patients with neurocysticercosis (Viola et al. 2011). Some of them have had subarachnoid hemorrhages

Table 7.1 Stroke syndromes due to neurocysticercosis

Clinical manifestations	Stroke subtype	Pathogenetic mechanism
Transient ischemic attacks	–	Narrowing of the intracranial internal carotid or basilar artery
Lacunar syndromes: ataxic hemiparesis, pure motor hemiparesis	Lacunar infarct in the internal capsule or the corona radiata	Inflammatory occlusion of small penetrating branches arising from the circle of Willis
Sensorimotor deficit, aphasia, signs of cortical dysfunction, coma	Large cerebral infarction involving the entire territory of the anterior or middle cerebral artery	Occlusion of major cerebral arteries due to atheroma-like deposits
Cognitive decline	Infarction in both frontal lobes	Occlusion of both anterior cerebral arteries
Top of the basilar syndrome, Parinaud's syndrome	Infarction involving the brainstem and thalamus	Inflammatory occlusion of penetrating branches of the basilar artery
Headache, vomiting, neck stiffness, coma	Subarachnoid hemorrhage	Rupture of a mycotic aneurysm
Headache, vomiting, focal neurological deficits	Parenchymal brain hemorrhage	Rupture of a small artery in the vicinity of a parenchymal brain cyst

related to the rupture of an inflammatory (mycotic) aneurysm induced by the inflammatory reaction that often surrounds subarachnoid cysticerci (Huang et al. 2000; Kim et al. 2005; Soto-Hernández et al. 1996). Some other patients have developed parenchymal brain hemorrhages that have been thought to be related to the damage of a small artery in the vicinity of a parenchymal brain cyst (Alarcón et al. 1992; Téllez-Zenteno et al. 2003).

7.3 Ventricular Neurocysticercosis

The main pathological substrate to explain clinical manifestations of intraventricular neurocysticercosis is the obstruction of CSF transit by the effect of parasites occluding the foramina of Monro, the cerebral aqueduct, or the foramina of Luschka and Magendie (Madrazo et al. 1983). Thus, patients often present with increased intracranial pressure of subacute onset that may be associated with episodes of loss of consciousness or even with sudden death due to acute hydrocephalus when freely floating cysts migrate from the lateral ventricles to the third ventricle or when fourth ventricle cysts move upwards and occlude the cerebral aqueduct (Llompart Pou et al. 2005; Sinha and Sharma 2012). Cysticerci located in the lateral ventricles may be associated with focal neurological signs when parasites compress adjacent structures such as the corticospinal tracts (Milenkovic et al. 1982). Cysticerci located in the fourth ventricle may be associated with signs of brainstem dysfunction due to compression of the floor of the fourth ventricle. A particular clinical form of presentation of fourth ventricle cysts is the Bruns' syndrome, characterized by episodic headache, neck stiffness, sudden positional vertigo, and loss of consciousness induced by rotatory movements of the head. While most patients rapidly recover after these episodes, some may die as the result of acute hydrocephalus (Jimenez Caballero et al. 2005). Granular ependymitis at the level of the cerebral aqueduct is also associated with signs and symptoms of increased intracranial pressure due to hydrocephalus. This condition is often associated with Parinaud's syndrome as the result of damage to the periaqueductal gray matter including the oculomotor nuclear complex (Keane 1982; Salazar et al. 1983).

7.4 Spinal Neurocysticercosis

Involvement of the spinal cord and the spinal canal in cysticercosis was considered to be extremely rare, accounting for less than 5 % of all neurocysticercosis cases (Isidro-Llorens et al. 1993). However, widespread use of MRI has improved the detection of this form of the disease (Gupta et al. 2009; Qi et al. 2011). Indeed, a recent study stresses the high frequency of spinal cysticerci among patients with intracranial subarachnoid neurocysticercosis, which can be as high as 61 % (Callacondo et al. 2012). Spinal cysticerci may be located within the spinal cord parenchyma or in the spinal subarachnoid space. Spinal cysts may be single or multiple, and, as occurs with other forms of neurocysticercosis, the clinical manifestations are mainly related to the number and location of the parasites.

Intramedullary cysts are most often located at the thoracic and lumbar segments of the spinal cord, although some patients with cervical cysts have been reported (Sheehan et al. 2002). Cysts produce motor deficits and sensory disturbances below the level of the lesion. Patients develop progressive spastic paraparesis and a sensory level including all sensory modalities; back pain and sphincter disturbances may be prominent complaints in some cases (Castillo et al. 1988; Qi et al. 2011; Sheehan et al. 2002). Intradural-extramedullary cysticercosis is more common than intramedullary cysts. These lesions may be single or may form clumps of multiple cysts extending along the entire spinal canal (Lim et al. 2010; Shin and Shin 2009). Clinical manifestations are characterized by a combination of radicular pain and motor deficits of subacute onset and progressive course (Callacondo et al. 2012; Gupta et al. 2009). Cervical leptomeningeal cysts may cause a compressive myelopathy with signs of upper motor neuron damage in the lower limbs associated with atrophy and fasciculations of hand muscles, mimicking amyotrophic lateral sclerosis (Kahn 1972). In other occasions, cervical leptomeningeal cysts may present as recurrent episodes of aseptic meningitis without focalizing signs (Boulos et al. 2010).

7.5 Other Locations of Intracranial Cysticerci

7.5.1 Intrasellar Neurocysticercosis

This rare form of neurocysticercosis presents with ophthalmological and endocrinological disturbances similar to those produced by pituitary tumors or craniopharyngiomas (Del Brutto et al. 1988b). In most cases, clinical manifestations are related to compression or even destruction of the pituitary stalk or the hypophysis itself by invasion of the sellar region by parasites (Arriada-Mendicoa et al. 2003). About 50 % of patients also present with subarachnoid or parenchymal brain cysticercosis which complicate even more the clinical course of the disease (Del Brutto and Del Brutto 2013b).

7.5.2 Subdural Neurocysticercosis

Cysticercosis of the subdural space is extremely rare, with only a couple of cases being reported so far (Feinberg and Valdivia 1984; Im et al. 2005).

Both patients presented with clinical and imaging evidence suggestive of chronic subdural hematomas; at surgery, lesions were found to be formed by multiple vesicular cysts of various sizes. In these cases, it is possible that parasites entered the subdural space from the contiguous subarachnoid space through the subarachnoid membrane.

7.5.3 Ocular and Orbital Cysticercosis

Cysticerci may be located in any part of the eye, including the anterior chamber, the lens, the vitreous, and the subretinal space (Madigubba et al. 2007). Subretinal cysts tend to locate near the macula causing progressive diminution of visual acuity or visual field defects. Cysticerci freely floating in the vitreous may also produce vision blurring that may be accompanied by a perception of movement of the cyst inside the eye. In addition, intraocular cysts may induce vitritis, uveitis, and endophthalmitis; the latter is the most severe complication of ocular cysticercosis and may lead to phthisis bulbi (Kruger-Leite et al. 1985). Extraocular cysticerci, mainly located in the conjunctiva and palpebra may cause unilateral proptosis, chronic conjunctivitis, and palpebral ptosis (Pushker et al. 2002). Retro-ocular intra-orbital cysticerci may compress the optic nerve with the subsequent development of diminution of visual acuity (Sudan et al. 2005) or may cause palpebral ptosis or diplopia due to direct involvement of extraocular muscles (Basu et al. 2009; Taksande et al. 2009).

7.6 Systemic Cysticercosis

Cysticerci may be located anywhere in the human body. While these locations must not be included in the definition of neurocysticercosis, a brief comment on these forms of the disease is important as they may coexist with intracranial involvement and may provide clues to the correct diagnosis in some cases (Del Brutto et al. 1996). Subcutaneous and striate muscle cysticerci are small and most often asymptomatic, although they may enlarge creating diagnostic confusion with neurofibromas, lipomas, or neoplasms (Sacchidanand et al. 2012). Massive cysticercal infection of striated muscles may eventually produces a syndrome of pseudohypertrophic myopathy with generalized weakness associated with painless and progressive muscle enlargement particularly affecting the calf muscles (Sawhney et al. 1976). Cysticercosis involvement of the peripheral nerves is rare; only a few patients have been reported. Cysts have been found in the median nerve causing paresthesias along the nerve's distribution and in the ulnar nerve causing weakness and wasting of hypothenar muscles (Nosanchuk et al. 1980; Sorasuchart et al. 1968). Finally, there are anecdotal reports of symptomatic visceral cysticercosis, including hepatic cysticercosis producing hepatomegaly with alterations of liver function tests (Sickel et al. 1995), myocardial cysticercosis presenting with symptoms of myocarditis and cardiac arrhythmias (Vieira de Melo et al. 2005), and thyroid cysts causing hypothyroidism (Gupta and Sodhani 2010).

References

Aditya GS, Mahadevan A, Santosh V, Chickabasaviah YT, Ashwathnarayanarao CB, Krishna SS (2004) Cysticercal chronic basal arachnoiditis with infarcts, mimicking tuberculous pathology in endemic areas. Neuropathology 24:320–325

Alarcón F, Hidalgo F, Moncayo J, Viñán I, Dueñas G (1992) Cerebral cysticercosis and stroke. Stroke 23:224–228

Arriada-Mendicoa N, Celis-López MA, Higuera-Calleja J, Corona-Vázquez T (2003) Imaging features of sellar cysticercosis. AJNR Am J Neuroradiol 24:1386–1389

Arruda WO (1991) Etiology of epilepsy. A prospective study of 210 cases. Arq Neuropsiquiatr 49: 251–254

Barinagarrementeria F, Del Brutto OH (1989) Lacunar syndrome due to neurocysticercosis. Arch Neurol 46:415–417

Basu S, Ramchandran U, Thapliyal A (2007) Clinical profile and outcome of pediatric neurocysticercosis: a study from Western Nepal. J Pediatr Neurol 5:45–52

Basu S, Muthusami S, Kumar A (2009) Ocular cysticercosis: an unusual cause of ptosis. Singapore Med J 50:e309

Bhalla A, Sood A, Sachdev A, Varma V (2008) Disseminated cysticercosis: a case report and review of the literature. J Med Case Reports 2:137

Bhattarai R, Budke CM, Carabin H, Proaño JV, Flores-Rivera J, Corona T, Cowan LD, Ivanek R, Snowden KF, Flisser A (2011) Quality of life in patients with neurocysticercosis in Mexico. Am J Trop Med Hyg 84:782–786

Blocher J, Schmutzhard E, Wilkins PP, Gupton PN, Schaffert M, Auer H, Gotwald T, Matuja W, Winkler AS (2011) A cross-sectional study of people with epilepsy and neurocysticercosis in Tanzania: clinical characteristics and diagnostic approaches. PLoS Negl Trop Dis 5:e1185

Bonametti AM, Baldy JLS, Bortoliero AL, de Maio CMD, Passos JN, Takata PK, de Pauli DS, Guimaraes JCA, Anzai ET, Elisbao MCM (1994) Neurocisticercose com quadro clínico inicial de meningite aguda. Rev Inst Med Trop Sao Paulo 36:27–32

Boulos MI, Aviv RI, Lee L (2010) Spinal neurocysticercosis manifesting as recurrent aseptic meningitis. Can J Neurol Sci 37:878–880

Callacondo D, García HH, Gonzales I, Escalante D, Nash TE (2012) High frequency of spinal involvement in patients with basal subarachnoid neurocysticercosis. Neurology 78:1394–1400

Cantú C, Barinagarrementeria F (1996) Cerebrovascular complications of neurocysticercosis. Clinical and neuroimaging spectrum. Arch Neurol 53:233–239

Carabin H, Ndimubanzi PC, Budke CM, Nguyen H, Qian Y, Cowan LD, Stoner JA, Rainwater E, Dickey M (2011) Clinical manifestations associated with neurocysticercosis: a systematic review. PLoS Negl Trop Dis 5(5):e1152

Castillo M, Quencer RM, Post MJD (1988) MR imaging of intramedullary spinal cysticercosis. AJNR Am J Neuroradiol 9:393–395
Chopra JS, Kaur U, Mahajan RC (1981) Cysticercosis and epilepsy: a clinical and serologic study. Trans R Soc Trop Med Hyg 75:518–520
Cosentino C, Velez M, Torres L, García HH (2002) Cysticercotic lesions in basal ganglia are common but clinically silent. Clin Neurol Neurosurg 104:57–60
Craig PS, Rogan MT, Allan JC (1996) Detection, screening and community epidemiology of taeniid cestode zoonoses: cystic echinococcosis, alveolar echinococcosis and neurocysticercosis. Adv Parasitol 38:169–250
Croker C, Redelings M, Reporter R, Sorvillo F, Mascola L, Wilkins P (2012) The impact of neurocysticercosis in California: a review of hospitalized cases. PLoS Negl Trop Dis 6(1):e1480
Cruz ME, Cruz I, Preux PM, Schantz P, Dumas M (1995) Headache and cysticercosis in Ecuador, South America. Headache 35:93–97
de Andrade DC, Rodrigues CL, Abraham R, Castro LHM, Livramento JA, Machado LR, Leire CC, Caramelli P (2010) Cognitive impairment and dementia in neurocysticercosis. A cross-sectional controlled study. Neurology 74:1288–1295
Del Brutto OH (1992) Cysticercosis and cerebrovascular disease: a review. J Neurol Neurosurg Psychiatry 55:252–254
Del Brutto OH (2012) Neurocysticercosis. Continuum (Minneap Minn) 18:1392–1416
Del Brutto OH (2013) Neurocysticercosis in infants and toddlers. Report of seven cases and review of published patients. Pediatr Neurol 48(6):432–435
Del Brutto OH, Campos X (2012) Massive neurocysticercosis: encephalitic versus non-encephalitic. Am J Trop Med Hyg 87:381
Del Brutto OH, Del Brutto VJ (2012a) Reduced percentage of neurocysticercosis cases among patients with late-onset Epilepsy in the new millennium. Clin Neurol Neurosurg 114: 1254–1256
Del Brutto OH, Del Brutto VJ (2012b) Calcified neurocysticercosis among patients with primary headache. Cephalalgia 32:250–254
Del Brutto OH, Del Brutto VJ (2013a) Isolated brainstem cysticercosis: a review. Clin Neurol Neurosurg 115:507–511
Del Brutto OH, Del Brutto VJ (2013b) Intrasellar cysticercosis: a systematic review. Acta Neurol Belg (in press)
Del Brutto OH, Noboa CA (1991) Late-onset epilepsy in Ecuador: aetiology and clinical features in 225 patients. J Trop Geogr Neurol 1:31–34
Del Brutto OH, Garcia E, Talamás O, Sotelo J (1988a) Sex-related severity of inflammation in parenchymal brain cysticercosis. Arch Intern Med 148:544–546
Del Brutto OH, Guevara J, Sotelo J (1988b) Intrasellar cysticercosis. J Neurosurg 69:58–60
Del Brutto OH, Santibáñez R, Noboa CA, Aguirre E, Díaz E, Alarcón TA (1992) Epilepsy due to neurocysticercosis: analysis of 203 patients. Neurology 42:389–392
Del Brutto OH, Wadia NH, Dumas M, Cruz M, Tsang VCW, Schantz PM (1996) Proposal of diagnostic criteria for human cysticercosis and neurocysticercosis. J Neurol Sci 142:1–6
Del Brutto OH, Santibáñez R, Idrovo L, Rodriguez S, Díaz-Calderón E, Navas C, Gilman RH, Cuesta F, Mosquera A, Gonzalez AE, Tsang VCW, García HH (2005) Epilepsy and neurocysticercosis in Atahualpa: a door-to-door survey in rural coastal Ecuador. Epilepsia 46:583–587
Feinberg WM, Valdivia FR (1984) Cysticercosis presenting as a subdural hematoma. Neurology 34:1112–1113
Feinstein A, Ron M, Wessely S (1990) Disappearing brain lesions, psychosis and epilepsy: a report of two cases. J Neurol Neurosurg Psychiatry 53:244–246
Finsterer J, Li M, Ramskogeler K, Auer H (2006) Chronic long-standing headache due to neurocysticercosis. Headache 46:523–524
Fleury A, Dessein A, Preux PM, Dumas M, Tapia G, Larralde C, Sciutto E (2004) Symptomatic human neurocysticercosis. Age, sex and exposure factors relating with disease heterogeneity. J Neurol 251:830–837

Fleury A, Escobar A, Fragoso G, Sciutto E, Larralde C (2010) Clinical heterogeneity of human neurocysticercosis results from complex interactions among parasite, host and environmental factors. Trans R Soc Trop Med Hyg 104:243–250

Fleury A, Carrillo-Mezo R, Flisser A, Sciutto E, Corona T (2011) Subarachnoid basal neurocysticercosis: a focus on the most severe form of the disease. Expert Rev Anti Infect Ther 9:123–133

Forlenza OV, Filho AHGV, Nobrega JPS, dos Ramos Machado L, de Barros NG, de Camargo CHP, da Silva MFG (1997) Psychiatric manifestations of neurocysticercosis: a study of 38 patients from a neurology clinic in Brazil. J Neurol Neurosurg Psychiatry 62:612–616

Gupta S, Sodhani P (2010) Clinically unsuspected thyroid involvement in cysticercosis: a case report. Acta Cytol 54(Suppl 5):853–856

Gupta S, Singh PK, Gupta B, Singh V, Azam A (2009) Isolated primary intradural extramedullary spinal neurocysticercosis: a case report and review of literature. Acta Neurol Taiwan 18: 187–192

Gupta RK, Awasthi R, Rathore RKS, Verma A, Sahoo P, Paliwal VK, Prasad KN, Pandey CM, Narayana PA (2012) Understanding epileptogenesis in calcified neurocysticercosis with perfusion MRI. Neurology 78:618–625

Huang PP, Choudhri HF, Jallo G, Miller DC (2000) Inflammatory aneurysm and neurocysticercosis: further evidence for a causal relationship? Case report. Neurosurgery 47:466–467

Im S-H, Park S-H, Oh DH, Kang B-S, Kwon O-K, Oh CW (2005) Subdural cysticercosis mimicking a chronic subdural hematoma. Case illustration. J Neurosurg 102:389

Isidro-Llorens A, Dachs F, Vidal J, Sarrias M (1993) Spinal cysticercosis. Case report and review. Paraplegia 31:128–130

Jimenez Caballero PE, Mollejo Villanueva M, Marsal Alonso C, Alvarez Tejerina A (2005) Síndrome de Bruns: descripción de un caso de neurocisticercosis con estudio anatomopatológico. Neurologia 20:87–89

Kahn P (1972) Cysticercosis of the central nervous system with amyotrophic lateral sclerosis: case report and review of the literature. J Neurol Neurosurg Psychiatry 35:81–87

Keane JR (1982) Neuro-ophthalmologic signs and symptoms of cysticercosis. Arch Ophthalmol 100:1445–1448

Keane JR (1993) Cysticercosis: unusual neuro-ophthalmologic signs. J Clin Neuroophthalmol 13:194–199

Kelesidis T, Tsiodras S (2012) Extraparenchymal neurocysticercosis in the United States. Am J Med Sci 344:79–82

Kim IY, Kim TS, Lee JH, Lee MC, Lee JK, Jung S (2005) Inflammatory aneurysm due to neurocysticercosis. J Clin Neurosci 12:585–588

Kruger-Leite E, Jalkh AE, Quiroz H, Schepens CL (1985) Intraocular cysticercosis. Am J Ophthalmol 99:252–257

Lim BC, Lee RS, Lim JS, Cho KY (2010) A case of neurocysticercosis at entire spinal level. J Korean Neurosurg Soc 48:371–374

Llompart Pou JA, Gené A, Ayestarán JI, Saus C (2005) Neurocysticercosis presenting as sudden death. Arch Neurochir (Wien) 147:785–786

Lobato RD, Lamas E, Portillo JM, Roger R, Esparza J, Rivas JJ, Muñoz MJ (1981) Hydrocephalus in cerebral cysticercosis. Pathogenic and therapeutic considerations. J Neurosurg 55:786–793

Madigubba S, Vishwanath K, Reddy G, Vemuganti GK (2007) Changing trends in ocular cysticercosis over two decades: an analysis of 118 surgically excised cysts. Indian J Med Microbiol 25:214–219

Madrazo I, García-Rentería JA, Sandoval M, López-Vega FJ (1983) Intraventricular cysticercosis. Neurosurgery 12:148–152

Marquez JM, Arauz A (2012) Cerebrovascular complications of neurocysticercosis. Neurologist 18:17–22

McCormick GF (1985) Cysticercosis—review of 230 patients. Bull Clin Neurosci 50:76–101

McCormick GF, Giannotta S, Zee C-S, Fisher M (1983) Carotid occlusion in cysticercosis. Neurology 33:1078–1080

Medina MT, Rosas E, Rubio-Donnadieu F, Sotelo J (1990) Neurocysticercosis as the main cause of late-onset epilepsy in Mexico. Arch Intern Med 150:325–327

Medina MT, Durón RM, Martínez L, Osorio JR, Estrada AL, Zúñiga C, Cartagena D, Collins JS, Holden KR (2005) Prevalence, incidence, and etiology of epilepsies in rural Honduras: the Salamá study. Epilepsia 46:124–131

Milenkovic Z, Penev G, Stojanovic D, Jovicic V, Antovic P (1982) Cysticercosis cerebri involving the lateral ventricle. Surg Neurol 18:94–96

Mishra D (2007) Cysticercosis headache: an important differential of childhood headache disorder in endemic countries. Headache 47:301–302

Monteiro de Almeida S, Gurjão SA (2011) Is the presence of depression independent from signs of disease activity in patients with neurocysticercosis? J Comm Health 36:693–697

Monteiro L, Almeida-Pinto J, Leite I, Xavier J, Correia M (1994) Cerebral cysticercus arteritis: five angiographic cases. Cerebrovasc Dis 4:125–133

Nash TE, Del Brutto OH, Butman JA, Corona T, Delgado-Escueta A, Duron RM, Evans CAW, Gilman RH, Gonzalez AE, Loeb JA, Medina MT, Pietsch-Escueta S, Pretell EJ, Takayanagui OM, Theodore W, Tsang VCW, García HH (2004) Calcific neurocysticercosis and epileptogenesis. Neurology 62:1934–1938

Ndimubanzi PC, Carabin H, Dudke CM, Nguyen H, Qian Y-J, Rainwater E, Dickey M, Reynolds S, Stoner JA (2010) A systematic review of the frequency of neurocysticercosis with a focus on people with epilepsy. PLoS Negl Trop Dis 4(11):e870

Nieto D (1956) Cysticercosis of the nervous system. Diagnosis by means of the spinal fluid complement fixation test. Neurology 6:725–738

Noboa CA (1992) Encefalitis cisticercosa. Rev Ecuat Neurol 1:61–71

Nosanchuk JS, Agostini JC, Georgi M, Posso M (1980) Pork tapeworm of cysticercus involving peripheral nerve. JAMA 244:2191–2192

Poon TP, Arida EJ, Tyschenko WP (1980) Cerebral cysticercosis with aqueductal obstruction. Case report. J Neurosurg 53:252–255

Preux P-M, Druet-Cabanac M (2005) Epidemiology and aetiology of epilepsy in sub-Saharan Africa. Lancet Neurol 4:21–31

Pushker N, Bajaj MS, Betharia SM (2002) Orbital and adnexal cysticercosis. Clin Experiment Ophthalmol 30:322–333

Qi B, Ge P, Yang H, Bi C, Li Y (2011) Spinal intramedullary cysticercosis: a case report and literature review. Int J Med Sci 8:420–423

Rajshekhar V (2000) Severe episodic headache as the sole presenting ictal event in patients with a solitary cysticercus granuloma. Acta Neurol Scand 102:44–46

Ramirez-Bermudez J, Higuera J, Sosa AL, Lopez-Meza E, Lopez-Gomez M, Corona T (2005) Is dementia reversible in patients with neurocysticercosis? J Neurol Neurosurg Psychiatry 76:1164–1166

Rangel R, Torres B, Del Brutto O, Sotelo J (1987) Cysticercotic encephalitis: a severe form in young females. Am J Trop Med Hyg 36:387–392

Rathore C, Radhakrishnan K (2012) What causes seizures in patients with calcified neurocysticercal lesions? Neurology 78:612–613

Rathore C, Thomas B, Kesavadas C, Radhakrishnan K (2012) Calcified neurocysticercosis and hippocampal sclerosis: potential dual pathology? Epilepsia 53:e60–e62

Revuelta-Gutierrez R, Valdéz-García J (1995) Cisticercosis como causa de hidrocefalia de presión normal (síndrome de Hakin-Adams). Arch Inst Nac Neurol Neurocir (Méx) 10:98–101

Rodrigues CL, de Andrade DC, Livramento JA, Machado LR, Abraham R, Massaroppe L, Lucato LT, Caramelli P (2012) Spectrum of cognitive impairment in neurocysticercosis. Differences according to disease phase. Neurology 78:861–866

Rodriguez-Carbajal J, Del Brutto OH, Penagos P, Huebe J, Escobar A (1989) Occlusion of the middle cerebral artery due to cysticercotic angiitis. Stroke 20:1095–1099

Sacchidanand S, Namitha P, Mallikarjuna M, Nataraj HV (2012) Disseminated cutaneous cysticercosis and neurocysticercosis: a rare occurrence. Indian Dermatol Online J 3:135–137

Salazar A, Sotelo J, Martinez H, Escobedo F (1983) Differential diagnosis between ventriculitis and fourth ventricle cyst in neurocysticercosis. J Neurosurg 59:660–663

Sawhney BB, Chopra JS, Banerji AK, Wahi PL (1976) Pseudohypertrophic myopathy in cysticercosis. Neurology 26:270–272

Shandera WX, White CA Jr, Chen JC, Díaz P, Armstrong R (1994) Cysticercosis in Houston, Texas. A report of 112 patients. Medicine 73:37–51

Sheehan JP, Sheehan JM, Lopes MB, Jane JA (2002) Intramedullary cervical spine cysticercosis. Acta Neurochir 144:1061–1063

Shin DA, Shin HC (2009) A case of extensive spinal cysticercosis involving the whole spinal canal in a patient with a history of cerebral cysticercosis. Yonsei Med J 50:582–584

Shriqui CL, Milette PC (1992) You drive me crazy: a case report of acute psychosis and neurocysticercosis. Can J Psychiatry 37:121–124

Sickel JZ, Fultz PJ, Penwarden B, Laczin J (1995) Hepatic cysticercosis. Report of an unusual case. J Clin Gastroenterol 20:160–163

Sinha S, Sharma BS (2012) Intraventricular neurocysticercosis: a review of current status and management issues. Br J Neurosurg 26:305–309

Sorasuchart A, Khunadorn N, Edmeads J (1968) Parasitic diseases of the nervous system in Thailand. Can Med Assoc J 98:859–864

Sotelo J (2011) Clinical manifestations, diagnosis, and treatment of neurocysticercosis. Clin Neurol Neurosci Rep 11:529–535

Sotelo J, Marin C (1987) Hydrocephalus secondary to cysticercotic arachnoiditis. A long-term follow-up review of 92 cases. J Neurosurg 66:686–689

Sotelo J, Guerrero V, Rubio F (1985) Neurocysticercosis: a new classification based on active and inactive forms. Arch Intern Med 145:442–445

Soto-Hernández JL, Gomez-Llata SA, Rojas-Echeverri LA, Texeira F, Romero V (1996) Subarachnoid hemorrhage secondary to a ruptured inflammatory aneurysm: a possible complication of neurocysticercosis: case report. Neurosurgery 38:197–199

Srinivas HV, Chandramukhi A (1992) Pseudopyogenic meningitis as a manifestation of neurocysticercosis – a case report. J Trop Geogr Neurol 2:140–141

Suástegui R, Gutierrez J, Ramos R, Bouchan S, Navarrete H, Ruiz J, Plascencia N, Jauri S, León C, Castillo V, Ojeda EA (2009) Características clínicas de la epilepsia de inicio tardío en México al principio del nuevo milenio: 455 casos. Rev Invest Clin 61:354–363

Sudan R, Muralidhar R, Sharma P (2005) Optic nerve cysticercosis: case report and review of current management. Orbit 24:159–162

Taksande B, Jajoo U, Yelwatkar S, Ashish J (2009) Unusual presentation of orbital cysticercosis – ptosis, diminution of vision and medial rectus weakness: a case report. Cases J 2:7025

Talukdar B, Saxena A, Popli VK, Choudhury V (2002) Neurocysticercosis in children: clinical characteristics and outcome. Ann Trop Pediatr 22:333–339

Téllez-Zenteno JF, Negrete-Pulido OR, Cantú C, Márquez C, Vega-Boada F, García-Ramos G (2003) Hemorrhagic stroke associated to neurocysticercosis. Neurologia 18:272–275

van As AD, Joubert J (1991) Neurocysticercosis in 578 black epileptic patients. S Afr Med J 80: 327–328

Vieira de Melo RM, Vieira de Melo Neto A, Correa LCL, Vieira de Melo Filho A, Filho A (2005) Restrictive cardiomyopathy due to myocardial cysticercosis. Arch Bras Cardiol 85:425–427

Villarán MV, Montano SM, Gonzalvez G, Moyano LM, Chero JC, Rodriguez S, Gonzalez AE, Pan W, Tsang VCW, Gilman RH, García HH (2009) Epilepsy and neurocysticercosis: an incidence study in a Peruvian rural population. Neuroepidemiology 33:25–31

Viola GM, White AC, Serpa JA Jr (2011) Hemorrhagic cerebrovascular events and neurocysticercosis: a case report and review of the literature. Am J Trop Med Hyg 84:402–405

Wadia NH (1995) Neurocysticercosis in Asia. Presented at the informal consultation of taeniasis/cysticercosis, Pan American Health Organization, Brasilia, Brazil, 23–25 Aug 1995

Wallin MT, Pretell EJ, Bustos JA, Caballero M, Alfaro M, Kane R, Wilken J, Sullivan C, Fratto T, García HH (2012) Cognitive changes and quality of life in neurocysticercosis: a longitudinal study. PLoS Negl Trop Dis 6:e1493

8 Diagnosis of Taeniasis and Cysticercosis

Despite the fact that human taeniasis and cysticercosis have a well-known etiologic agent, diagnosis of these conditions may be a challenge. In the case of taeniasis, most infected individuals are asymptomatic or may develop vague complaints and, as previously described, defining a typical syndrome of cysticercosis is not possible. Such clinical pleomorphism is often complicated by the scarcity of pathognomonic findings or the lack of reliability of many of the available diagnostic tests. Here, we describe the results of the most accepted complementary exams used for the diagnosis of the disease complex taeniasis/cysticercosis.

8.1 Diagnosis of Taeniasis

Humans carrying the adult *Taenia solium* in the intestine are the source of cysticercosis infection and should be ruled out in individuals with neurocysticercosis and their household contacts, particularly in young patients or those with massive infestations (Asnis et al. 2009; Del Brutto and Del Brutto 2013a; García and Del Brutto 1999; Gilman et al. 2000). The chances of finding a tapeworm carrier among household contacts of a neurocysticercosis patient are logically smaller in older patients with only calcified lesions, since they may have acquired the infection many years ago and the life span of the adult *Taenia solium* in tapeworm seems to be below 5 years (Allan et al. 1996a; Huisa et al. 2005).

8.1.1 Microscopy

The traditional parasitological test, i.e., microscopically observation of Taenia eggs, is of poor use for the diagnosis of *T. solium* taeniasis, even if concentration methods are used. In the best hands, the sensitivity of coproparasitologic microscopy stands below 70 % (García et al. 2003a). Reasons for this include the variable rate of proglottid and egg excretion, that most eggs remain in the proglottids, and the slower growth rate compared to other tapeworms. To partially leverage the poor sensitivity, concentration methods

O.H. Del Brutto, H.H. García, *Cysticercosis of the Human Nervous System*, DOI 10.1007/978-3-642-39022-7_8, © Springer-Verlag Berlin Heidelberg 2014 should be used, preferably using sedimentation; larger sample volumes need to be concentrated, and repeated testing (three different samples) is recommended. On the other hand, the specificity of human stool microscopy for taeniasis is very high. Appearance of Taenia egg is quite characteristic, with a thick radiate cover and hooks (embryonic hooks, not cysticercal hooks). However, *Taenia solium* eggs are extremely similar than those of *Taenia saginata*; thus a diagnosis of *Taenia sp.* infection should only be reported. The occurrence of morphological and staining differences between the two tapeworms has recently been suggested, but this is not absolute (Jimenez et al. 2010).

8.1.2 Stool Antigen Detection

Allan et al. (1990, 1993) applied capture ELISA assays to detect *Taenia solium* antigens in stool samples, as previously reported for *Taenia hydatigena* and other parasites. This coproantigen-detection ELISA greatly improved the diagnostic sensitivity for human taeniasis. In field surveys, coproantigen detection finds between 1.5 and 2.5 times more taeniasis cases than the classic microscopic stool examination (Allan et al. 1996b; García et al. 2003b). The usual coproantigen ELISA assay uses a polyclonal antibody against the adult tapeworm stage to detect tapeworm antigen in stool supernatant. A 5 % PBS-formaldehyde solution is used as the diluent and fixative to avoid biosafety risks.

The coproantigen-detection assay uses a very low cutoff value calculated from negative samples. Low positive results are frequently nonspecific, but strongly positive samples usually correspond to tapeworm carriers. Samples from *Taenia saginata* carriers tend to have very low positive results, and only a small proportion of the general population have a consistently positive stool antigen result, likely due to a cross-reaction with rheumatoid factor (Yolken and Stopa 1979). Effective treatment with niclosamide or praziquantel rapidly decreases stool antigen levels, and 2 weeks after treatment almost all cured patients are coproantigen negative; thus, a positive test after 2 weeks or 1 month of treatment marks treatment failure (much before the tapeworm begins to release eggs again) (Bustos et al. 2012).

8.1.3 DNA-Based Assays

The use of PCR assays in the diagnosis of human taeniasis has been described by several groups. These assays can discriminate the species of *Taenia* and are highly reliable if tapeworm material is available. This is not usually the case, and so far the sensitivity of PCR assays in stool samples is not yet defined (Gonzalez et al. 2000; Mayta et al. 2000; Rodríguez et al. 2012).

8.2 Imaging Diagnosis of Neurocysticercosis

Introduction of modern neuroimaging techniques has greatly increased the diagnostic accuracy for neurocysticercosis by providing objective evidence on the number and topography of lesions, as well as on the activity of the disease, including the degree of the host's inflammatory reaction against this parasite (Rodriguez-Carbajal and Boleaga-Durán 1982). These imaging methods have replaced previously used radiological procedures such as plain roentgenograms, pneumoencephalograms, ventriculography, conventional myelography, and angiography that currently have only limited or historical importance (Arana and Asenjo 1945; Dorfsman 1963; Trelles and Trelles 1978).

8.2.1 Old Imaging Diagnostic Methods

Small, single, or multiple intracranial calcifications are common on plain X-ray films of the skull in patients with neurocysticercosis (Dixon and Hargreaves 1944; Stepien 1962). These calcifications are typically small and are characterized by an eccentric calcified dot representing the scolex, surrounded by a partially calcified sphere that represents the body of the cystic larva (Santín and Vargas 1966). These characteristics were once considered useful for differentiating calcifications related to cysticercosis from those arising as the result of other infections such as toxoplasmosis or tuberculosis (Rodriguez-Carbajal and Boleaga-Durán 1982). Another finding on plain X-ray films of the skull was the erosion of the sella turcica as a reflection of chronic intracranial hypertension, usually related to hydrocephalus (Cárdenas y Cárdenas 1962). Soft-tissue roentgenograms are still useful for the detection of cysticerci in muscles. Parasites may appear as small punctate or as elliptical, cigar-shaped calcifications (McCormick 1985). These lesions are more frequent in calf and thigh muscles, and in areas where modern neuroimaging diagnostic tests are not readily available, this is a useful radiological sign for the diagnosis of human cysticercosis and provides indirect evidence favoring the diagnosis of neurocysticercosis in patients with neurological complaints (Dumas et al. 1990; Preux et al. 1996).

Pneumoencephalograms and ventriculography were the imaging methods of choice for the diagnosis of intraventricular cysticercosis by identifying filling defects within the ventricular system that corresponded to ventricular cysts. These methods were also of value for the evaluation of the size of the ventricular cavities in patients with suspected hydrocephalus (Bickerstaff et al. 1952; Milenkovic et al. 1982). The main problem was their lack of specificity, since any intraventricular space-occupying lesion may produce similar filling defects.

Myelography was of great diagnostic value in patients with suspected spinal leptomeningeal cysticercosis by showing multiple filling defects in the column of contrast material corresponding to the cysts (Firemark 1978). Using myelography, leptomeningeal cysts may be seen freely mobile within the spinal subarachnoid space and may change their position according to movements of the patient in the exploration table (Kim and Weinberg 1985). However, myelography is not specific in patients with intramedullary cysts of the spinal cord since the method only visualizes partial or complete blockage in the column of contrast material at the level of the lesion, a finding that may also be observed in other diseases of the spinal cord (Akiguchi et al. 1979).

Angiography has been used for the diagnosis of neurocysticercosis since the pioneer description of Moniz et al. (1932). Early reports were most often concerned with the displacement of cerebral arteries caused by hydrocephalus and did not assign a major diagnostic value to this method (Rocca and Monteagudo 1966). More recently, reports dealing with cerebrovascular complications of neurocysticercosis described in detail the spectrum of angiographic changes observed in this parasitic disease, including segmental narrowing or occlusion of small- and middle-size intracranial vessels in patients with cerebral infarctions (Barinagarrementeria and Del Brutto 1989; Levy et al. 1995; Monteiro et al. 1994; Rodriguez-Carbajal et al. 1989) as well as mycotic aneurysms in patients with subarachnoid hemorrhage (Soto-Hernández et al. 1996). Nowadays, conventional angiography has been largely replaced by magnetic resonance angiography and by CT angiograms for the diagnosis of intracranial vascular lesions in patients with neurocysticercosis (Del Brutto 2008).

8.2.2 Computed Tomography (CT) and Magnetic Resonance Imaging (MRI)

Introduction of CT really improved the diagnosis and understanding of neurocysticercosis as it allowed, for the very first time, accurate preoperative visualization of cysticerci as well as proper identification of the number and topography of parasites and their stages of viability or involution (Carbajal et al. 1977; Bentson et al. 1977). CT also allowed the development of clinical classifications of neurocysticercosis based on the topography of lesions and the activity of the disease (Sotelo et al. 1985) and was important for determining the rational therapeutic approach for the different forms of the disease. Then, the advent of MRI provided a series of imaging advantages over CT, including the possibility of multi-planar (axial, coronal, and sagittal) reconstruction of images, the capacity to visualize the posterior fossa without bone artifacts, and a contrast resolution that far exceeds that of CT (Martinez et al. 1989; Suss et al. 1986). A major shortcoming of MRI, however, is failure to detect small calcifications which—in many cases—represent the sole imaging evidence of the disease. Because of this major limitation of MRI, CT remains the best screening neuroimaging procedure for patients with suspected neurocysticercosis. Other disadvantages of MRI include the scarcity of the equipment in poor regions where cysticercosis is endemic and the high costs of the exam.

8.2.2.1 Parenchymal Neurocysticercosis

The stage of viability and involution of parenchymal brain cysticerci determines their appearance on CT and MRI. Viable (vesicular) cysticerci appear as small and rounded cystic lesions that are well demarcated from the surrounding brain parenchyma. There is little or no perilesional swelling and no abnormal enhancement after contrast medium administration. Many vesicular cysts have in their interior a brightening nodule representing the scolex, giving the lesions a pathognomonic "hole-with-dot" appearance (Fig. 8.1). When the infection is massive, as in the so-called heavy nonencephalitic form of neurocysticercosis (García and Del Brutto 1999), the brain looks like a "Swiss cheese" (Fig. 8.2) or even as a "starry night" (Fig. 8.3). During the first stage of involution (colloidal stage), cysticerci appear as ill-defined lesions surrounded by edema, and most of them show an abnormal nodular or ring pattern of

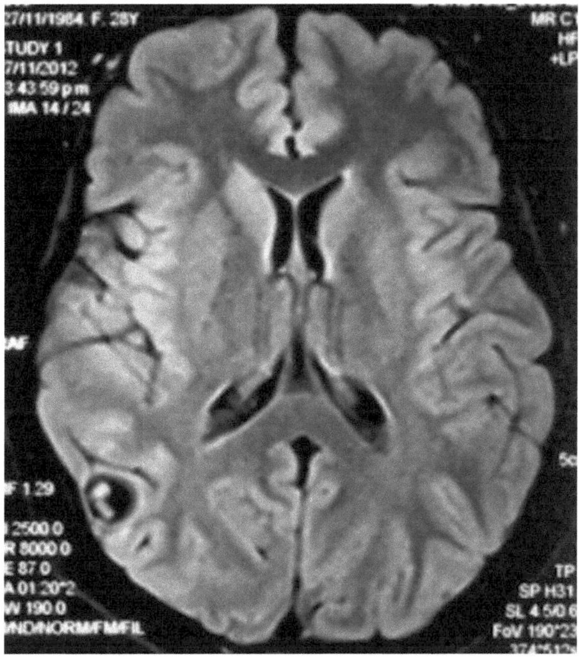

Fig. 8.1 Fluid attenuation inversion recovery MRI showing single parenchymal brain cysticerci in the vesicular stage located in the right parietal lobe. The scolex is clearly visualized within the cyst

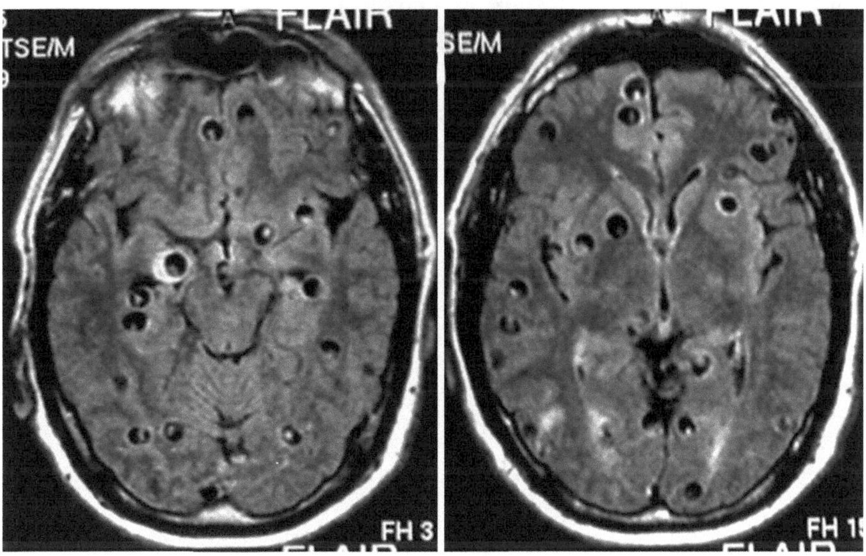

Fig. 8.2 Contrast-enhanced MRI of a patient with heavy non-encephalitic parenchymal brain cysticercosis giving the brain a "Swiss cheese" appearance

enhancement after contrast medium administration; the scolex is seldom visualized in colloidal cysticerci using CT or conventional MRI sequences (Fig. 8.4). Therefore, the presence of one or two ring-enhancing lesion in the brain parenchyma may represent a diagnostic challenge, as other infections or even neoplasms may present with

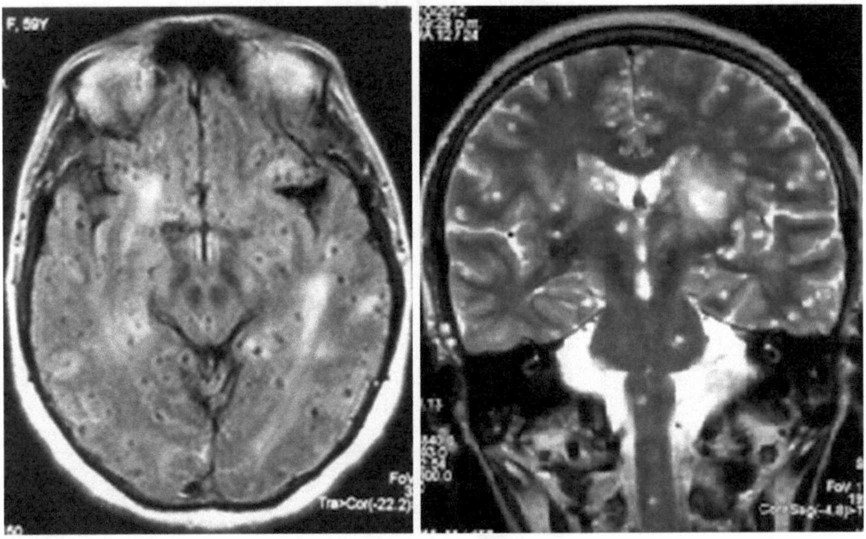

Fig. 8.3 Fluid attenuation inversion recovery (*left*) and T2-weighted (*right*) MRI of a patient with massive infection of the brain parenchyma by small cysticerci, giving the so-called starry night effect

Fig. 8.4 Contrast-enhanced MRI showing two colloidal parenchymal brain cysticerci (Reproduced from Del Brutto (2012a), ©Oscar H. Del Brutto)

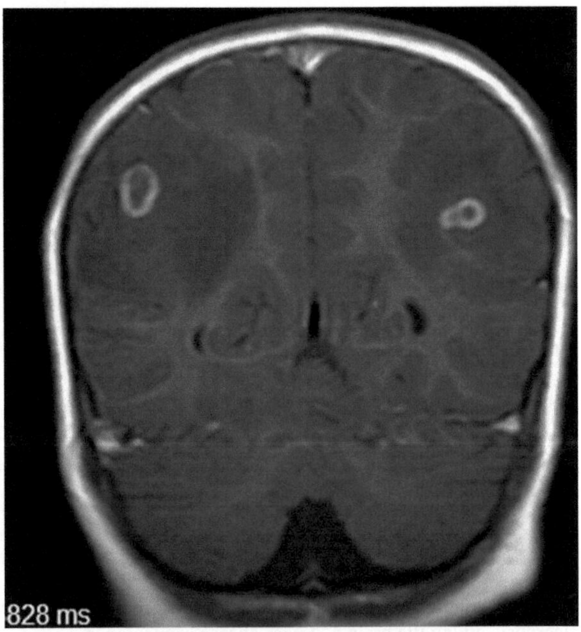

similar neuroimaging findings. In such cases, the practice of diffusion-weighted images and apparent diffusion coefficient maps facilitates the diagnosis by allowing the recognition of the scolex (do Amaral et al. 2005) (Fig. 8.5).

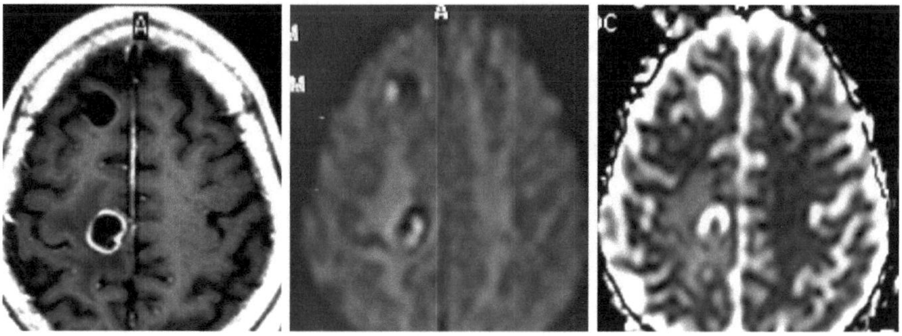

Fig. 8.5 MRI of a patient with colloidal parenchymal brain cysticerci. Scolices are not visualized on contrast-enhancing imaging (*left*), but are well seen in diffusion-weighted images (*center*) and in the apparent diffusion coefficient map (*right*) (Reproduced with permission from: Del Brutto (2012b))

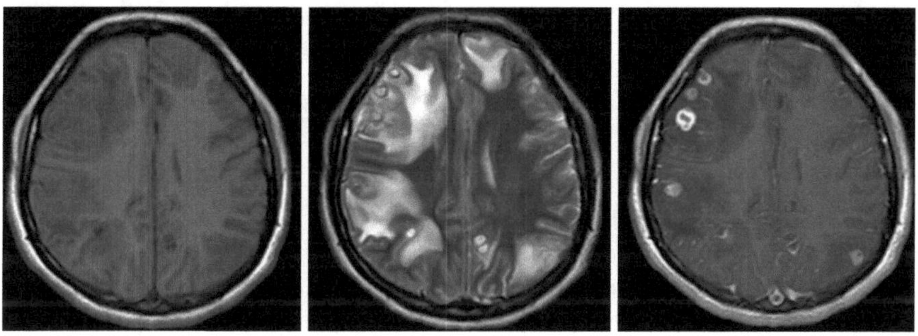

Fig. 8.6 MRI of a patient with cysticercotic encephalitis. T1-weighted image showing diffuse brain edema and collapse of the ventricular system (*left*), T2-weighted imaging showing multiple colloidal cysticerci surrounded by edema (*center*) and contrast-enhanced imaging showing ring-enhancing lesions (*right*) (Reproduced with permission from: Del Brutto (2012b))

A particular neuroimaging pattern of parenchymal neurocysticercosis in the colloidal stage is that of the so-called cysticercotic encephalitis. In this severe form of the disease, both CT and MRI show diffuse or multifocal brain edema and collapse of the ventricular system without midline shift (Del Brutto and Campos 2012). In such cases, multiple small nodular or ring-enhancing lesions appear disseminated through the brain parenchyma after contrast medium administration (Fig. 8.6).

Parenchymal brain cysticerci may also appear as slightly hyperdense/hyperintense nodular-enhancing lesions surrounded or not by edema or perilesional gliosis. This pattern corresponds to the granular stage of cysticerci which, as previously noted, represents degenerated cysticerci. Calcified cysticerci appear, on CT, as small hyperdense nodules not surrounded by edema or abnormal contrast enhancement. As noted before, the sensitivity of conventional MRI sequences for the detection of calcified lesions is not as good as that of CT (Fig. 8.7). There is some recent evidence, however, that use of susceptibility-weighted images may enhance the visualization of calcifications by MRI (Wu et al. 2009).

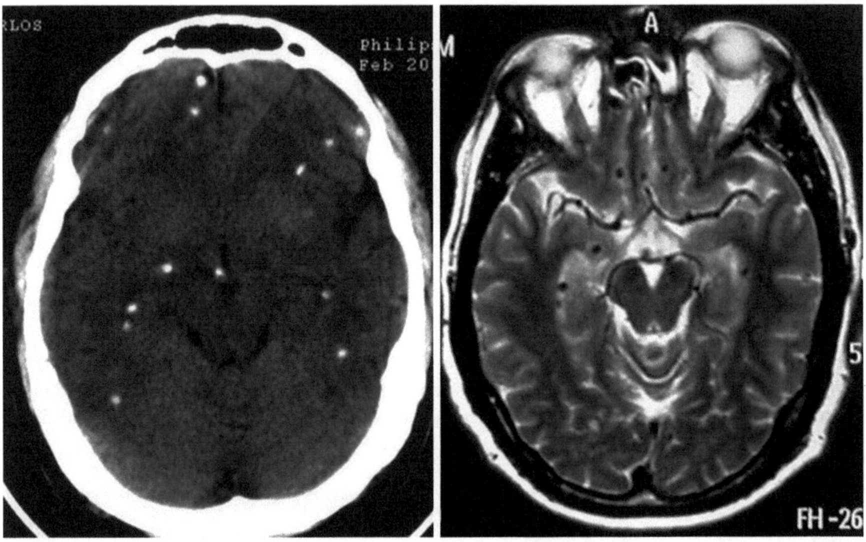

Fig. 8.7 Contrast-enhanced CT (*left*) and T2-weighted MRI (*right*) showing multiple parenchymal brain calcifications, which are more easily discernible with CT than with MRI

8.2.2.2 Subarachnoid Neurocysticercosis

Subarachnoid cysts located within cortical sulci may present with similar neuroimaging findings than those described for parenchymal brain cysts or may grow to distort the cerebral anatomy and compress neighboring structures (Fig. 8.8). Lesions located within the Sylvian fissures or at the basal CSF cisterns may have a multilobulated appearance and often attain a large size, displacing neighboring structures and behaving as space-occupying mass lesions (Fig. 8.9). Hydrocephalus, related to inflammatory occlusion of Luschka and Magendie foramina, may also be seen. The fibrous arachnoiditis that is responsible for the development of hydrocephalus is seen on neuroimaging studies as abnormal leptomeningeal enhancement at the base of the brain (Fleury et al. 2011; Kimura-Hayama et al. 2010).

Cysticerci located in the sellar region are rare. Those lesions may be confined to the intrasellar compartment or may be associated with hydrocephalus or subarachnoid cysts at the suprasellar cistern (Del Brutto et al. 1988). According to a recent review, MRI is better than CT to visualize those cysts (Del Brutto and Del Brutto 2013b).

Cerebrovascular complications of neurocysticercosis are well visualized with modern neuroimaging techniques. In patients with cysticercosis-related cerebral infarctions, the association of subarachnoid cystic lesions or abnormal enhancement of basal leptomeninges suggests the correct diagnosis (Del Brutto 2008). Magnetic resonance angiography is a valuable noninvasive imaging modality to demonstrate narrowing or occlusion of intracranial arteries in patients with subarachnoid neurocysticercosis (Fig. 8.10).

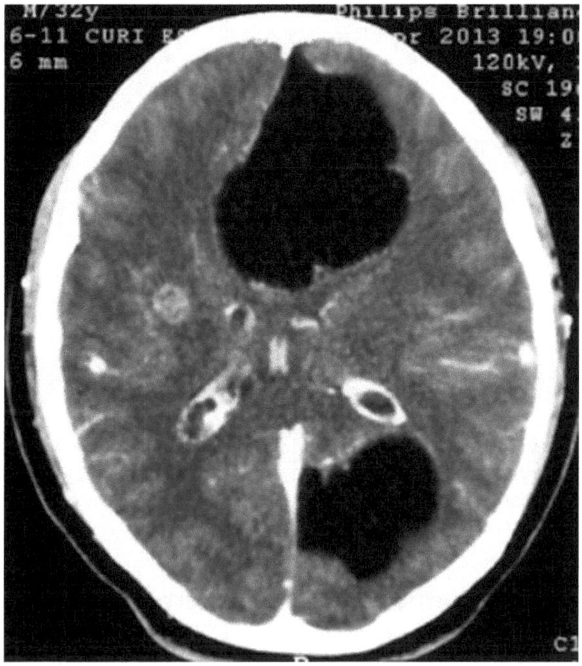

Fig. 8.8 Contrast-enhanced CT showing two large subarachnoid cysticerci arising from cortical sulci that distort brain anatomy and compress neighboring structures

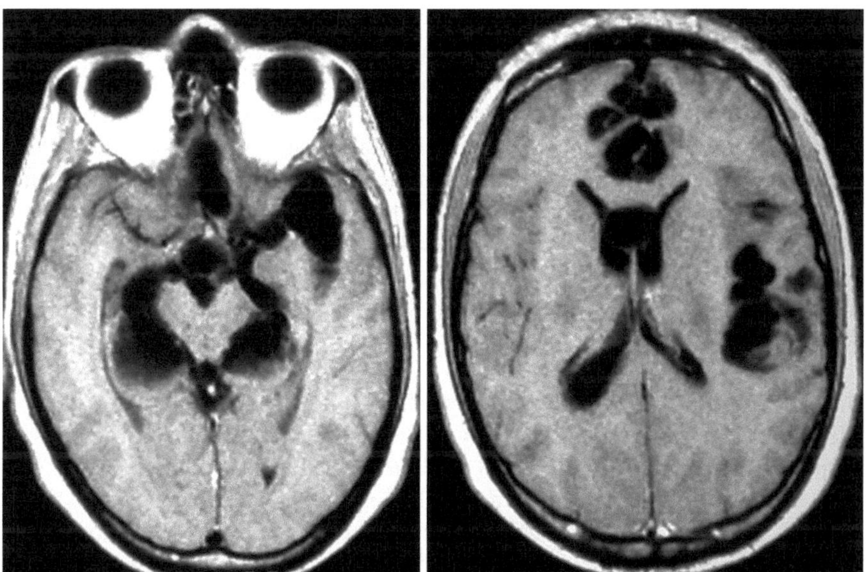

Fig. 8.9 T1-weighted MRI showing clumps of subarachnoid cysticerci involving the left Sylvian fissure and the basal CSF cisterns surrounding the midbrain

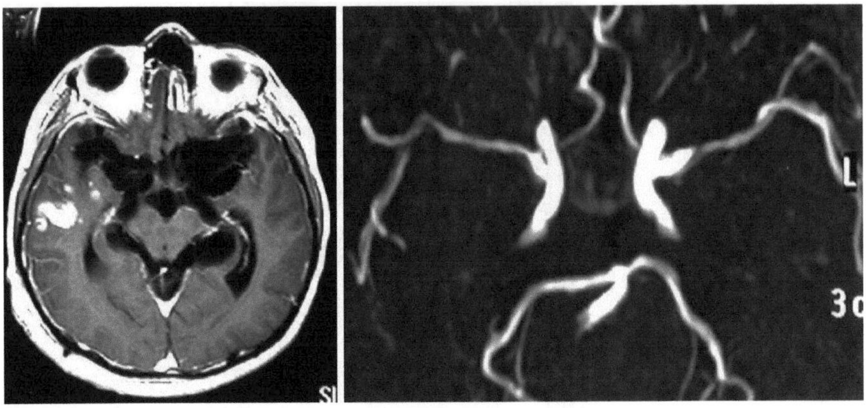

Fig. 8.10 Contrast-enhanced MRI of a patient presenting with an acute stroke syndrome, showing large clumps of cysticerci within the basal subarachnoid cisterns and a recent infarction in the right temporal lobe (*left*) and MRA showing multiple stenosis of major intracranial arteries arising from the circle of Willis (*right*)

8.2.2.3 Ventricular Neurocysticercosis

Ventricular cysticerci appear on CT as hypodense lesions that cause asymmetric obstructive hydrocephalus. Most often, ventricular cysts cannot be visualized directly as they are isodense with CSF. In contrast, ventricular cysts are readily visualized on MRI because the signal properties of the cystic fluid, the scolex, or the vesicular wall differ from those of the CSF (Kimura-Hayama et al. 2010) (Fig. 8.11). Cyst mobility within the ventricular cavities in response to movements of the head, the so-called ventricular migration sign, facilitates the diagnosis of ventricular cysticercosis in some cases. In other cases, parasitic membranes or inflammation of the ependymal lining (ventriculitis) occludes Monro's foramina and causes asymmetric internal hydrocephalus, which is most often noticed after the placement of a ventricular shunt, as the lateral ventricle contralateral to the shunt remains dilated after the derivative procedure. A rare finding in patients with ventricular cysticercosis is the so-called double compartment hydrocephalus (DeFeo et al. 1975), in which the fourth ventricle is isolated from the rest of ventricular cavities and from the subarachnoid space due to simultaneous occlusion of the cerebral aqueduct and the foramina of Luschka and Magendie.

8.2.2.4 Spinal Cord Neurocysticercosis

MRI has become the imaging modality of choice for the evaluation of patients with suspected cysticercosis of the spinal cord or the spinal subarachnoid space. MRI allows precise localization of intramedullary cysticerci (Homans et al. 2001; Qi et al. 2011). Those cysts are usually single and most often located at the lower segments of the thoracic spinal cord, although they may also be seen at the cervical or lumbosacral levels. Most of these cysts appear as ring- or nodular-enhancing lesions, and the scolex can be visualized in a sizable subset of cases (references). Leptomeningeal cysts are also easily identified with MRI (Fig. 8.12). These lesions

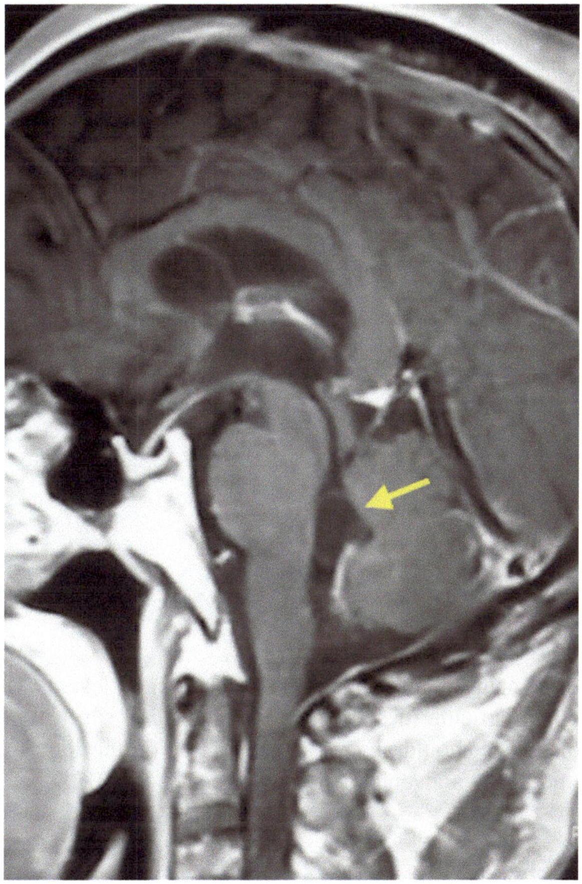

Fig. 8.11 Contrast-enhanced MRI showing cysticercus within the fourth ventricle (*arrow*). The vesicular wall is clearly distinguished from the CSF

may be freely mobile within the spinal subarachnoid space and may change their position during the exam according to movements of the patient in the exploration table. A recent study stresses the high frequency of spinal cysticerci among patients with intracranial subarachnoid neurocysticercosis (Callacondo et al. 2012). It is likely that this study will change the diagnostic approach of patients with intracranial subarachnoid neurocysticercosis, since all of them will probably need MRI investigation of the spine to detect hidden lesions.

8.3 Immunological Diagnosis of Neurocysticercosis

An ideal immunodiagnostic test for human neurocysticercosis should be 100 % sensitive (detecting all cases), 100 % specific (ruling out all cases without the disease), and provides information on the presence or not of living parasites as well as an idea of the parasite burden. Existing tests are far from this goal, and even if this could be achieved, immunodiagnosis would still not provide important information available

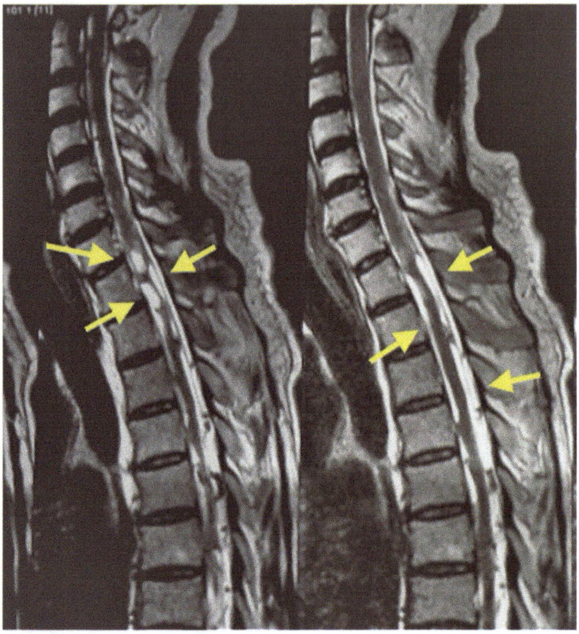

Fig. 8.12 T2-weighted MRI showing multiple spinal subarachnoid cysticerci (*arrows*)

from neuroimaging such as the location of lesions, degree of inflammation, and other particularities affecting the medical or surgical management. Thus, immunodiagnosis should always be interpreted on the light of clinical and neuroimaging findings in a given patient (Rodríguez et al. 2012).

8.3.1 Antibody Detection

More than 100 years ago, it was reported the detection of specific antibody reactions to cysticercal antigens by means of the complement fixation text in pig serum (Moses 1909) and in human CSF (Weinberg 1911). Along the years, a series of antibody detection techniques have been applied to the diagnosis of cysticercosis including immunofluorescence, hemagglutination, enzyme-linked immunosorbent assay (ELISA), and lately the EITB (immunoblot) (Arambulo et al. 1978; Diaz et al. 1992; Diwan et al. 1982; Larralde et al. 1986; Tsang et al. 1989).

The ELISA technique, described in 1971, did overcome previous tests because of its better performance and quantitative results. It was applied to the diagnosis of neurocysticercosis by the use of a crude cysticercal antigen since 1978 (Arambulo et al. 1978) and widely adopted a few years later. ELISA-based assays seem to perform better in CSF than in serum due to less likelihood of cross-reactions. Indeed, cross-reactions with other helminths are common, particularly with cestodes including *Echinococcus* spp. and *Hymenolepis nana*. In 1989, the EITB using a group of seven well-defined lentil-lectin purified glycoprotein antigens provided a much more sensitive and specific assay and became the test of choice for the diagnosis of cysticercosis (Tsang et al. 1989). Strong antibody reactions suggest severe infections and are infrequent in asymptomatic populations in community-based studies (Corona et al. 1986; Zini et al. 1990).

Initial evaluations of cysticercosis serology were blurred because of poor characterization of positive cases (parenchymal/extraparenchymal, numbers of viable lesions, etc.) and by the use of different antigens and test protocols, among other reasons. The EITB assay is a complicated technique, and the production of lentil-lectin purified glycoprotein antigens requires fresh cysts which are difficult to obtain. Recent efforts have been oriented to produce synthetic or recombinant antigens for use in EITB or simpler techniques. So far, none of these tests has performed at the levels of the native EITB (Greene et al. 2000; Hancock et al. 2004, 2006).

Some caveats exist in antibody-based serodiagnosis. The presence of circulating antibodies does not discriminate viable from nonviable infections, since antibodies can result from exposure only, or from self-resolved infections. For this reason, a positive antibody result should be interpreted in the context of the studied population. In endemic regions, up to 20 % or 25 % of the general population may be antibody positive on EITB (García et al. 2003a; Tsang and Wilson 1995). In community settings most seropositive individuals are asymptomatic. Residual parenchymal brain calcifications are commonly found in these individuals and much more rarely, viable cysticerci (Del Brutto et al. 2005; Medina et al. 2005; Montano et al. 2005). The intensity of the reaction (number of reactive antibody bands) provides a gross indicator, with weak reactions in the general population and stronger reactions in clinical cases (García et al. 2003b). In individuals with a single brain lesion, even the EITB test may miss 30–40 % of cases, most likely due to very low levels of circulating antibody in these mild infections.

8.3.2 Antigen Detection

Detection of parasitic antigens demonstrates the presence of viable parasites and thus active infection. Antigen levels usually show a direct correlation with the numbers of cysts and extension of the parasite lesions. Two assays using monoclonal antibodies (originally developed against *Taenia saginata*) to detect circulating antigens of *Taenia solium* have been described, both with very similar performance (Brandt et al. 1992; Harrison et al. 1989). These assays were initially reported for use in CSF and later proved to be useful for the diagnosis of cysticercosis in serum samples.

8.3.3 Other Immune Diagnostic Tests

DNA-based PCR assays in CSF have been applied to the diagnosis of NCC with encouraging initial results (Almeida et al. 2006; Hernandez et al. 2005; Michelet et al. 2011). One of these series suggested higher sensitivity in patients with extraparenchymal NCC, and one reported problems with the specificity of the test. Further, systematic evaluations of their sensitivity should clarify their potential use for diagnosis. Also, a cellular-based immune assay, the lymphocyte transformation test was applied to neurocysticercosis in 2008, reporting a sensitivity of 94 and 96 % specificity (Prasad et al. 2008). This test is based on memory T cells and demonstrates previous exposure of the host to the antigen.

8.3.4 Antigen Versus Antibody Testing

Antigen-capture assays are—in general terms—less sensitive than antibody assays, which use antibody production by the host's immune system as a response intensifier. Very high antigen levels should thus raise the suspicion of extraparenchymal or massive parenchymal neurocysticercosis. Negative results are to be expected in patients with only a few parenchymal brain lesions. On the other hand, these assays seem to be highly specific in humans and do not seem to cross-react with other infections. Here, as in any other ELISA test, the intensity of the reaction should be taken into account in a similar way as described above for the EITB, and reactions close to the cutoff level should be interpreted with caution. Detection of circulating antigen seems very promising as a tool to monitor the evolution of patients, particularly in cases of extraparenchymal neurocysticercosis (Bobes et al. 2006; Fleury et al. 2013; Zamora et al. 2005).

The major advantage of antibody detection over antigen detection is a much higher sensitivity, which makes antibody-detecting assays the tool of choice for serological diagnosis. On the other hand, circulating antibody does not discriminate between active and inactive infections and thus its clinical utility is restricted to etiological confirmation, and since antibodies persist for long time, antibody detection assays are impractical to monitor disease evolution. Antigen levels give an idea of the infection burden and severity of infection (Fleury et al. 2003; García et al. 2000), and their negativization is strongly suggestive of complete parasite resolution.

8.3.5 Interpretation of a Positive Immune Diagnostic Test

Immunodiagnosis, either antigen or antibody detection, is not specific for central nervous system infection, neither can they provide specific data on the characteristics of the lesions. It follows that antigen or antibody assay results should be interpreted in the context of an individual patient or population, and neuroimaging is required to guide therapeutic decisions.

Antibody and antigen seroprevalence issues were previously discussed in the chapter of epidemiology in endemic regions and here we will refer only to their use in the clinical context. An attending physician caring for people from cysticercosis-endemic regions should have in mind that these regions have a background prevalence of seropositivity which can reach 20 % or more of the general population.

However, these usually are weak reactions. In clinical settings, a weak positive antibody test is only to be expected in patients with a single viable lesion or calcified neurocysticercosis. In these cases, a weak result is compatible with the clinical presentation although it could still result from the background population positivity levels. In most other cases antibody reactions should be clearly positive, and these clearly positive tests in the presence of neurological manifestations have a very high predictive value to confirm the diagnosis.

If neuroimaging is not available, the use of serology to diagnose cysticercosis becomes a matter of discussion and depends on whether etiological management (antiparasitic drugs or surgery) could be given in blind. As will be discussed latter on this book, we do not recommend to consider initiating blind antiparasitic treatment for neurocysticercosis, based on the following reasons: (a) the initial steps of antiparasitic treatment do not result in immediate symptom relief, conversely, a transitory exacerbation in symptoms can occur and requires appropriate steroid and symptomatic treatment; (b) theoretically, degeneration of a cyst in a delicate location, i.e., the ventricular system, could cause serious complications or even be lethal; and (c) symptomatic treatment is key and it is mostly independent of the etiology of the symptoms. Alternatively, a positive immune diagnostic test should be considered as a criterion for patient referral for imaging diagnosis and etiological case management. It seems counterintuitive to assume a smaller need of referral for other symptomatic conditions different to neurocysticercosis.

8.4 Other Diagnostic Tests for Human Cysticercosis

Some hematological abnormalities, i.e., peripheral eosinophilia, have been described in patients with neurocysticercosis; however, these are inconstant and have no diagnostic strength (Loo and Braude 1982). In addition, abnormalities in the cytochemical composition of CSF have been reported in a sizable proportion of patients with neurocysticercosis (McCormick et al. 1982). These changes are nonspecific and do not differentiate neurocysticercosis from other infections of the brain or the spinal cord. Moreover, a normal CSF examination does not rule out the diagnosis of neurocysticercosis. The most common finding is a moderate mononuclear pleocytosis (cell counts rarely exceeding 300/mm^3) (Bittencourt et al. 1990). A nonspecific, and often transient, eosinophilia may be noticed in almost 60 % of the patients who have pleocytosis (Johnson 1989). Proteins may be raised in the range of 50–300 mg/dl, although CSF protein levels as high as 1,600 mg/dl have been reported (McCormick et al. 1982). CSF glucose levels are usually within the normal range despite active meningeal disease. Indeed, this finding has been considered as a useful differential diagnostic clue between cysticercotic and other infections of the nervous system, i.e., neurotuberculosis, which are usually associated with low CSF glucose levels. Nevertheless, very low glucose levels (<10 mg/dl) have been reported in patients with neurocysticercosis and have been associated with poor prognosis (McCormick et al. 1982).

8.5 Diagnostic Criteria for Neurocysticercosis

As described throughout this chapter, the diagnosis of neurocysticercosis can be a challenge. In 1996, the first attempt to settle a set of diagnostic criteria for human cysticercosis was published. It was based on the objective evaluation of clinical, radiological, immunological, and epidemiological data of patients (Del Brutto et al. 1996). After some years of experience, it was considered that a set of criteria exclusively devoted to the diagnosis of neurocysticercosis would be more comprehensible than the initial one (Del Brutto et al. 2001). As in the 1996 publication, the revised criteria included four categories of diagnosis—absolute, major, minor, and epidemiologic—stratified on the basis of their individual strength. Absolute criteria allow unequivocal diagnosis of neurocysticercosis, major criteria strongly suggest the diagnosis but could not be used alone to confirm the diagnosis, minor criteria are frequent but nonspecific manifestations of the disease, and epidemiologic criteria refer to circumstantial evidence favoring the diagnosis. Interpretation of these criteria permits two degrees of diagnostic certainty: (a) *definitive diagnosis*, in patients who have one absolute criterion or in those who have two major plus one minor and one epidemiologic criteria, and (b) *probable diagnosis*, in patients who have one major plus two minor criteria, in those who have one major plus one minor and one epidemiologic criteria, and in those who have three minor plus one epidemiologic criteria (Table 8.1). More than 10 years after its publication, the 2001 set of diagnostic criteria has become a useful tool for the diagnosis of neurocysticercosis in patients living in endemic as well as in non-endemic countries, avoiding over- and underdiagnosis of the disease in both field studies and the hospital setting.

Table 8.1 Diagnostic criteria for neurocysticercosis

Diagnostic criteria
Absolute
Histologic demonstration of the parasite from biopsy of a brain or spinal cord lesion
Evidence of cystic lesions showing the scolex on neuroimaging studies
Direct visualization of subretinal parasites by fundoscopic examination
Major
Evidence of lesions highly suggestive of neurocysticercosis on neuroimaging studies
Positive serum immunoblot for the detection of anticysticercal antibodies
Resolution of intracranial cystic lesions after therapy with albendazole or praziquantel
Spontaneous resolution of small single enhancing lesions
Minor
Evidence of lesions compatible with neurocysticercosis on neuroimaging studies
Presence of clinical manifestations suggestive of neurocysticercosis

Table 8.1 (continued)

Positive CSF ELISA for detection of anticysticercal antibodies or cysticercal antigens
Evidence of cysticercosis outside the central nervous system
Epidemiologic
Individuals coming from or living in an area where cysticercosis is endemic
History of travel to disease-endemic areas
Evidence of a household contact with *T. solium* infection
Degrees of diagnostic certainty
Definitive
Presence of one absolute criterion
Presence of two major plus one minor and one epidemiologic criteria
Probable
Presence of one major plus two minor criteria
Presence of one major plus one minor and one epidemiologic criteria
Presence of three minor plus one epidemiologic criteria

Adapted from Del Brutto et al. (2001)

References

Akiguchi I, Fujiwara T, Matsuyama H, Muranaka H, Kameyama M (1979) Intramedullary spinal cysticercosis. Neurology 29:1531–1534

Allan JC, Avila G, Garcia Noval J, Flisser A, Craig PS (1990) Immunodiagnosis of taeniasis by coproantigen detection. Parasitology 101 Pt3:473–477

Allan JC, Mencos F, Garcia-Noval J, Sarti E, Flisser A, Wang Y, Liu D, Craig PS (1993) Dipstick dot ELISA for the detection of Taenia coproantigens in humans. Parasitology 107(Pt 1): 79–85

Allan JC, Velasquez-Tohorn M, Garcia-Noval J, Torres-Alvarez R, Yurrita P, Fletes C, de Mata F, Soto de Alfaro H, Craig PS (1996a) Epidemiology of intestinal taeniasis in four, rural, Guatemalan communities. Ann Trop Med Parasitol 90:157–165

Allan JC, Velasquez-Tohorn M, Torres-Alvarez R, Yurrita P, Garcia-Noval J (1996b) Field trial of the coproantigen-based diagnosis of *Taenia solium* taeniasis by enzyme-linked immunosorbent assay. Am J Trop Med Hyg 54:352–356

Almeida CR, Ojopi EP, Nunez CM, Machado LR, Takayanagui OM, Livramento JA, Abraham R, Gattaz WF, Vaz AJ, Dias-Neto E (2006) *Taenia solium* DNA is present in the cerebrospinal fluid of neurocysticercosis patients and can be used for diagnosis. Eur Arch Psychiatry Clin Neurosci 256:307–310

Arambulo PV 3rd, Wall KW, Bullock S, Kagan IG (1978) Serodiagnosis of human cysticercosis by microplate enzyme-linked immunospecific assay (ELISA). Acta Trop 35:63–67

Arana R, Asenjo A (1945) Ventriculographic diagnosis of cysticercosis of the posterior fossa. J Neurosurg 2:181–190

Asnis D, Kazakov J, Toronjadze T, Bern C, García HH, McAuliffe I, Bishop H, Lee L, Grossmann R, Garcia MA, Di John D (2009) Neurocysticercosis in the infant of a pregnant mother with a tapeworm. Am J Trop Med Hyg 81:449–451

Barinagarrementeria F, Del Brutto OH (1989) Lacunar syndrome due to neurocysticercosis. Arch Neurol 46:415–417

Bentson JR, Wilson GH, Helmer E, Winter J (1977) Computed tomography in intracranial cysticercosis. J Comput Assist Tomogr 1:464–471

Bickerstaff ER, Cloake PCP, Hughes B, Smith WT (1952) The racemose form of cerebral cysticercosis. Brain 75:1–18

Bittencourt PRM, Costa JM, Oliveira TV, Gracia CM, Gorz AM, Mazer S (1990) Clinical, radiological and cerebrospinal fluid presentation of neurocysticercosis: a prospective study. Arq Neuropsiquiatr 48:286–295

Bobes RJ, Hernandez M, Marquez C, Fragoso G, Garcia E, Parkhouse RM, Harrison LJ, Sciutto E, Fleury A (2006) Subarachnoid and intraventricular human neurocysticercosis: application of an antigen detection assay for the diagnosis and follow-up. Am J Trop Med Hyg 81: 449–451

Brandt JR, Geerts S, De Deken R, Kumar V, Ceulemans F, Brijs L, Falla N (1992) A monoclonal antibody-based ELISA for the detection of circulating excretory-secretory antigens in *Taenia saginata* cysticercosis. Int J Parasitol 22:471–477

Bustos JA, Rodriguez S, Jimenez JA, Moyano LM, Castillo Y, Ayvar V, Allan JC, Craig PS, Gonzalez AE, Gilman RH, Tsang VC, García HH (2012) Detection of *Taenia solium* taeniasis coproantigen is an early indicator of treatment failure for taeniasis. Clin Vaccine Immunol 19:570–573

Callacondo D, García HH, Gonzales I, Escalante D, Nash TE (2012) High frequency of spinal involvement in patients with basal subarachnoid neurocysticercosis. Neurology 78: 1394–1400

Carbajal JR, Palacios E, Azar-Kia B, Churchill R (1977) Radiology of cysticercosis of the central nervous system including computed tomography. Radiology 125:127–131

Cárdenas y Cárdenas J (1962) Cysticercosis of the nervous system. II. Pathologic and radiologic findings. J Neurosurg 19:635–640

Corona T, Pascoe D, Gonzalez-Barranco D, Abad P, Landa L, Estañol B (1986) Anticysticercus antibodies in serum and cerebrospinal fluid in patients with cerebral cysticercosis. J Neurol Neurosurg Psychiatry 49:1044–1049

DeFeo D, Foltz EL, Hamilton AE (1975) Double compartment hydrocephalus in a patient with cysticercotic meningitis. Surg Neurol 4:247–251

Del Brutto OH (2008) Stroke and vasculitis in patients with cysticercosis. In: Caplan LR (ed) Uncommon causes of stroke. Cambridge University Press, New York

Del Brutto OH (2012a) Neurocysticercosis in a 2-year-old boy infected at home. Pathog Global Health 106:122–123

Del Brutto OH (2012b) Neurocysticercosis. Continuum lifelong learning in neurology. J Am Acad Neurol 18:1392–1416

Del Brutto OH, Campos X (2012) Massive neurocysticercosis: encephalitic versus non-encephalitic. Am J Trop Med Hyg 87:381

Del Brutto OH, Del Brutto VJ (2013a) Neurocysticercosis in infants and toddlers. Report of seven cases and review of published patients. Pediatr Neurol 48(6):432–435

Del Brutto OH, Del Brutto VJ (2013b) Intrasellar cysticercosis – a systematic review. Acta Neurol Belg (in press)

Del Brutto OH, Guevara J, Sotelo J (1988) Intrasellar cysticercosis. J Neurosurg 69:58–60

Del Brutto OH, Wadia NH, Dumas M, Cruz M, Tsang VCW, Schantz PM (1996) Proposal of diagnostic criteria for human cysticercosis and neurocysticercosis. J Neurol Sci 142:1–6

Del Brutto OH, Rajshekhar V, White AC Jr, Tsang VC, Nash TE, Takayanagui OM, Schantz PM, Evans CA, Flisser A, Correa D, Botero D, Allan JC, Sarti E, Gonzalez AE, Gilman RH, García HH (2001) Proposed diagnostic criteria for neurocysticercosis. Neurology 57:177–183

Del Brutto OH, Santibáñez R, Idrovo L, Rodriguez S, Díaz-Calderón E, Navas C, Gilman RH, Cuesta F, Mosquera A, Gonzalez AE, Tsang VCW, García HH (2005) Epilepsy and neurocysticercosis in Atahualpa: a door-to-door survey in rural coastal Ecuador. Epilepsia 46: 583–587

Diaz JF, Verastegui M, Gilman RH, Tsang VC, Pilcher JB, Gallo C, García HH, Torres P, Montenegro T, Miranda E (1992) Immunodiagnosis of human cysticercosis (*Taenia solium*): a field comparison of an antibody-enzyme-linked immunosorbent assay (ELISA), an antigen-ELISA, and an enzyme-linked immunoelectrotransfer blot (EITB) assay in Peru. The Cysticercosis Working Group in Peru (CWG). Am J Trop Med Hyg 46:610–615

Diwan AR, Coker-Vann M, Brown P, Subianto DB, Yolken R, Desowitz R, Escobar A, Gibbs CJ Jr, Gajdusek DC (1982) Enzyme-linked immunosorbent assay (ELISA) for the detection of antibody to cysticerci of *Taenia solium*. Am J Trop Med Hyg 31:364–369

Dixon HBF, Hargreaves WH (1944) Cysticercosis (*Taenia solium*). A further ten years' clinical study covering 284 cases. Q J Med 13:107–121

do Amaral LLF, Ferreira RM, da Rocha AJ, Ferreira NP (2005) Neurocysticercosis. Evaluation with advances magnetic resonance techniques and atypical forms. Top Magn Reson Imaging 16:127–144

Dorfsman J (1963) The radiologic aspects of cerebral cysticercosis. Acta Radiol Diagn (Stockh) 1:836–842

Dumas M, Grunitzky K, Belo M, Dabis F, Daniau M, Bouteille B, Kassankogno Y, Catanzano G, Alexandre MP (1990) Cysticercose et neurocysticercose: enquête épidémiologique dans le nord du Togo. Bull Soc Path Exot 83:263–274

Firemark HM (1978) Spinal cysticercosis. Arch Neurol 35:250–251

Fleury A, Hernandez M, Fragoso G, Parkhouse RM, Harrison LJ, Sciutto E (2003) Detection of secreted cysticercal antigen: a useful tool in the diagnosis of inflammatory neurocysticercosis. Trans R Soc Trop Med Hyg 97:542–546

Fleury A, Carrillo-Mezo R, Flisser A, Sciutto E, Corona T (2011) Subarachnoid basal neurocysticercosis: a focus on the most severe form of the disease. Expert Rev Anti Infect Ther 9: 123–133

Fleury A, Garcia E, Hernandez M, Carrillo R, Govenzensky T, Fragoso G, Sciutto E, Harrison LJ, Parkhouse RM (2013) Neurocysticercosis: HP10 antigen detection is useful for the follow-up of the severe patients. PLoS Negl Trop Dis 7(3):e2096

García HH, Del Brutto OH (1999) Heavy nonencephalitic cerebral cysticercosis in tapeworm carriers. The Cysticercosis Working Group in Perú. Neurology 53:1582–1584

García HH, Parkhouse RM, Gilman RH, Montenegro T, Bernal T, Martinez SM, Gonzalez AE, Tsang VC, Harrison LJ (2000) Serum antigen detection in the diagnosis, treatment, and follow-up of neurocysticercosis patients. Trans R Soc Trop Med Hyg 94:673–676

García HH, Gilman RH, Gonzalez AE, Verastegui M, Rodriguez S, Gavidia C, Tsang VC, Falcon N, Lescano AG, Moulton LH, Bernal T, Tovar M (2003a) Hyperendemic human and porcine *Taenia solium* infection in Peru. Am J Trop Med Hyg 68:268–275

García HH, Gonzalez AE, Gilman RH (2003b) Diagnosis, treatment and control of *Taenia solium* cysticercosis. Curr Opin Infect Dis 16:411–419

Gilman RH, Del Brutto OH, García HH, Martínez M (2000) Prevalence of taeniosis among patients with neurocysticercosis is related to severity of infection. The Cysticercosis Working Group in Peru. Neurology 55:1062

Gonzalez AE, Montero E, Harrison LJ, Parkhouse RM, Garate T (2000) Differential diagnosis of *Taenia saginata* and *Taenia solium* infection by PCR. J Clin Microbiol 38:737–744

Greene RM, Hancock K, Wilkins PP, Tsang VC (2000) *Taenia solium*: molecular cloning and serologic evaluation of 14- and 18-kDA related, diagnostic antigens. J Parasitol 86:1001–1007

Hancock K, Pattabhi S, Greene RM, Yushak ML, Khan A, Priest JW, Levine MZ, Tsang VC (2004) Characterization and cloning of GP50, a *Taenia solium* antigen diagnostic for cysticercosis. Mol Biochem Parasitol 133:115–124

Hancock K, Pattabhi S, Whitfield FW, Yushak ML, Lane WS, García HH, Gonzalez AE, Gilman RH, Tsang VC (2006) Characterization and cloning of T24, a *Taenia solium* antigen diagnostic for cysticercosis. Mol Biochem Parasitol 147:109–117

Harrison LJ, Joshua GW, Wright SH, Parkhouse RM (1989) Specific detection of circulating surface/secreted glycoproteins of viable cysticerci in *Taenia saginata* cysticercosis. Parasite Immunol 11:351–370

Hernandez M, Gonzalez LM, Fleury A, Saenz B, Parkhouse RM, Harrison LJ, Garate T, Sciutto E (2005) Neurocysticercosis: detection of *Taenia solium* DNA in human cerebrospinal fluid using a semi-nested PCR based on HDP2. Ann Trop Med Parasitol 102:317–323

Homans J, Khoo L, Chen T, Commins DL, Ahmed J, Kovacs A (2001) Spinal intramedullary cysticercosis in a five-year-old child: case report and review of the literature. Pediatr Infect Dis J 20:904–908

Huisa BN, Benacho LA, Rodriguez S, Bustos JA, Gilman RH, Tsang VC, Gonzalez AE, García HH (2005) Taeniasis and cysticercosis in housemaids working in affluent neighborhoods in Lima, Peru. Am J Trop Med Hyg 73:496–500

Jimenez JA, Rodriguez S, Moyano LM, Castillo Y, García HH (2010) Differentiating Taenia eggs found in human stools: does Ziehl-Neelsen staining help? Trop Med Int Health 15:1077–1081

Johnson LN (1989) Neurocysticercosis. Arch Neurol 46:842–843

Kim KS, Weinberg PE (1985) Spinal cysticercosis. Surg Neurol 24:80–82

Kimura-Hayama ET, Higuera JA, Corona-Cedillo R, Chávez-Macías L, Perochena A, Quiroz-Rojas LY, Rodríguez-Carbajal J, Criales JL (2010) Neurocysticercosis: radiologic-pathologic correlation. Radiographics 30:705–719

Larralde C, Laclette JP, Owen CS, Madrazo I, Sandoval M, Bojalil R, Sciutto E, Contreras L, Arzate J, Diaz ML (1986) Reliable serology of *Taenia solium* cysticercosis with antigens from cyst vesicular fluid: ELISA and hemagglutination tests. Am J Trop Med Hyg 35:965–973

Levy AS, Lillehei KO, Rubinstein D, Stears JC (1995) Subarachnoid neurocysticercosis with occlusion of the major intracranial arteries: case report. Neurosurgery 36:183–188

Loo L, Braude A (1982) Cerebral cysticercosis in San Diego. A report of 23 cases and a review of the literature. Medicine 61:341–359

Martinez HR, Rangel-Guerra R, Elizondo G, Gonzalez J, Todd LE, Ancer J, Prakash SS (1989) MR imaging in neurocysticercosis: a study of 56 cases. AJNR Am J Neuroradiol 10: 1011–1019

Mayta H, Talley A, Gilman RH, Jimenez J, Verastegui M, Ruiz M, García HH, Gonzalez AE (2000) Differentiating *Taenia solium* and *Taenia saginata* infections by simple hematoxylin-eosin staining and PCR-restriction enzyme analysis. J Clin Microbiol 38:133–137

McCormick GF (1985) Cysticercosis – review of 230 patients. Bull Clin Neurosci 50:76–101

McCormick GF, Zee C-S, Heiden J (1982) Cysticercosis cerebri. Review of 127 cases. Arch Neurol 39:534–539

Medina MT, Durón RM, Martínez L, Osorio JR, Estrada AL, Zúñiga C, Cartagena D, Collins JS, Holden KR (2005) Prevalence, incidence, and etiology of epilepsies in rural Honduras: the Salamá study. Epilepsia 46:124–131

Michelet L, Fleury A, Sciutto E, Kendjo E, Fragoso G, Paris L, Bouteille B (2011) Human cysticercosis: comparison of different diagnostic tests using cerebrospinal fluid. J Clin Microbiol 49:195–200

Milenkovic Z, Penev G, Stojanovic D, Jovicic V, Antovic P (1982) Cysticercosis cerebri involving the lateral ventricle. Surg Neurol 18:94–96

Moniz E, Loff R, Pacheco L (1932) Sur le diagnostic de la cysticercose cérébrale. A propos de deux cas. Encéphale 27:42–53

Montano SM, Villaran MV, Ylquimiche L, Figueroa JJ, Rodriguez S, Bautista CT, Gonzalez AE, Tsang VC, Gilman RH, García HH (2005) Neurocysticercosis: association between seizures, serology, and brain CT in rural Peru. Neurology 65:229–233

Monteiro L, Almeida-Pinto J, Leite I, Xavier J, Correia M (1994) Cerebral cysticercus arteritis: five angiographic cases. Cerebrovasc Dis 4:125–133

Prasad A, Prasad KN, Yadav A, Gupta RK, Pradhan S, Jha S, Tripathi M, Husain M (2008) Lymphocyte transformation test: a new method for diagnosis of neurocysticercosis. Diag Microbiol Inf Dis 61:198–202

Preux PM, Melaku Z, Druet-Cabanac M, Avode G, Grunitzky EK, Bouteille B, Cruz M, Dumas M (1996) Cysticercosis and neurocysticercosis in Africa: current status. Neurol Infect Epidemiol 1:63–68

Qi B, Ge P, Yang H, Li Y (2011) Spinal intramedullary cysticercosis: a case report and literature review. Int J Med Sci 8:420–423

Rocca E, Monteagudo E (1966) An angiographic study of neurocysticercosis. Int J Surg 46:130–141

Rodríguez S, Wilkins P, Dorny P (2012) Immunological and molecular diagnosis of cysticercosis. Pathog Global Health 106:286–298

Rodriguez-Carbajal J, Boleaga-Durán B (1982) Neuroradiology of human cysticercosis. In: Flisser A, Willms K, Laclete JP, Larralde C, Ridaura C, Beltrán F (eds) Cysticercosis: present state of knowledge and perspectives. Academic, New York

Rodriguez-Carbajal J, Del Brutto OH, Penagos P, Huebe J, Escobar A (1989) Occlusion of the middle cerebral artery due to cysticercotic angiitis. Stroke 20:1095–1099

Santín G, Vargas J (1966) Roentgen study of cysticercosis of the central nervous system. Radiology 86:520–528

Sotelo J, Guerrero V, Rubio F (1985) Neurocysticercosis: a new classification based on active and inactive forms. Arch Intern Med 145:442–445

Soto-Hernández JL, Gomez-Llata S, Rojas-Echeverri LA, Texeira F, Romero V (1996) Subarachnoid hemorrhage secondary to a ruptured inflammatory aneurysm: a possible manifestation of neurocysticercosis: case report. Neurosurgery 38:197–200

Stepien L (1962) Cerebral cysticercosis in Poland. Clinical symptoms and operative results in 132 cases. J Neurosurg 19:505–513

Suss RA, Maravilla KR, Thompson J (1986) MR imaging of intracranial cysticercosis: comparison with CT and anatomopathologic features. AJNR Am J Neuroradiol 7:235–242

Trelles JO, Trelles L (1978) Parasitic diseases and tropical neurology. In: Vinken PJ, Bruyn GQ (eds) Handbook of clinical neurology, vol 35. North Holland, Amsterdam

Tsang V, Wilson M (1995) *Taenia solium* cysticercosis, an under-recognized but serious public health problem. Parasitol Today 11:124–126

Tsang VC, Brand JA, Boyer AE (1989) An enzyme-linked immunotransfer blot assay and glycoprotein antigens for diagnosing human cysticercosis (*Taenia solium*). J Infect Dis 159:50–59

Wu Z, Mittal S, Kish K, Yu Y, Hu J, Haacke EM (2009) Identification of calcification with MRI using susceptibility-weighted imaging: a case study. J Mag Res Imaging 29:177–182

Yolken RH, Stopa PJ (1979) Analysis of nonspecific reactions in enzyme-linked immunosorbent assay testing for human rotavirus. J Clin Microbiol 10:703–707

Zamora H, Castillo Y, García HH (2005) Drop in antigen levels following successful treatment of subarachnoid neurocysticercosis. Am J Trop Med Hyg 73:S41

Zini D, Farrell VJ, Wadee AA (1990) The relationship of antibody levels to the clinical spectrum of human neurocysticercosis. J Neurol Neurosurg Psychiatry 53:656–661

Management of Neurocysticercosis 9

As detailed in the preceding chapters, clinical manifestations of neurocysticercosis are directly related to the localization of parasites within the nervous system, and result from inflammation, mass effects, blockage of the CSF transit, fibrosis, or vasculitis (García and Del Brutto 2005). Inflammation arises from the attack of the host's immune system to the parasite and is usually found by the time a cyst degenerates. Inflammation is not necessarily a continuous progressive process and occasionally an inflamed cyst may return to a quiescent stage. Moreover, inflammation may also present around old, calcified lesions (Nash et al. 1999). Mass effect occurs when large cysts or cyst clumps located in the subarachnoid space compress neighboring structures, although it may also result from smaller lesions exerting pressure to an adjacent eloquent cerebral area (Proaño et al. 2001). Blockage of the CSF transit is usually caused by intraventricular or subarachnoid lesions or due to chronic scarring resulting from residual fibrosis (Estañol et al. 1983). More rarely, local or remote damage to small- and medium-sized intracranial arteries by a surrounding subarachnoid parasite, or by the inflammatory reaction developed in the subarachnoid space, may result in the occurrence of an ischemic stroke with the resulting clinical manifestations (Del Brutto 2008).

A first line of management should target the presenting symptoms of a given patient and pathogenetic mechanisms involved in their occurrence. Therefore, appropriate institution and optimization of symptomatic therapy—antiepileptic drugs (AEDs), analgesics, anti-inflammatory drugs, anti-edema agents, or surgery—should always precede the use of cysticidal agents (Nash and García 2011). This concept is of particular importance since during the initial days or weeks after the onset of cysticidal drug therapy, symptoms will not improve and could even get worse; this may be dangerous in a given patient with intracranial hypertension or poorly controlled seizures.

As noted, neurocysticercosis is a pleomorphic disease that causes several neurological syndromes and pathological lesions. Therefore, a unique therapeutic scheme cannot be useful in every patient. Characterization of the disease in terms of cysts' viability, degree of the host's immune response to the parasites, and location of the lesions are important for a rational therapy (García et al. 2002; Nash et al. 2006a, b).

O.H. Del Brutto, H.H. García, *Cysticercosis of the Human Nervous System*,
DOI 10.1007/978-3-642-39022-7_9,

Table 9.1 General guidelines for therapy of neurocysticercosis

Parenchymal neurocysticercosis
Vesicular cysts: *Single cyst:* Albendazole 15 mg/kg/day for 3 days or praziquantel 30 mg/kg in three divided doses every 2 h. Corticosteroids rarely needed. AED for seizures *Mild to moderate infections:* Albendazole 15 mg/kg/day for 1 week or praziquantel 50 mg/kg/day for 15 days. Corticosteroids may be used when necessary. AED for seizures *Heavy infections:* Albendazole 15 mg/kg/day for 1 week (repeated cycles of albendazole may be needed). Corticosteroids are mandatory before, during, and after therapy. AED for seizures
Colloidal cysts: *Single cyst:* Albendazole 15 mg/kg/day for 3 days or praziquantel 30 mg/kg in three divided doses every 2 h. Corticosteroids may be used when necessary. AED for seizures *Mild to moderate infections:* Albendazole 15 mg/kg/day for 1 week. Corticosteroids are usually needed before and during therapy. AED for seizures *Cysticercotic encephalitis:* Cysticidal drugs are contraindicated. Corticosteroids and osmotic diuretics to reduce brain swelling. AED for seizures. Decompressive craniectomies in refractory cases
Granular and calcified cysticerci: *Single or multiple:* No need for cysticidal drug therapy. AED for seizures. Corticosteroids for patients with recurrent seizures and perilesional edema surrounding calcifications
Extraparenchymal neurocysticercosis
Small cysts over convexity of cerebral hemispheres: *Single or multiple:* Albendazole 15 mg/kg/day for 1 week. Corticosteroids may be used when necessary. AED for seizures
Large cysts in Sylvian fissures or basal CSF cisterns: *Racemose cysticercus:* Albendazole 15–30 mg/kg/day for 15–30 days (repeated cycles of albendazole may be needed). Corticosteroids are mandatory before, during, and after therapy
Other forms of extraparenchymal neurocysticercosis: *Hydrocephalus:* No need for cysticidal drugs therapy. Ventricular shunt. Continuous corticosteroid administration (50 mg three times a week for up to 2 years) may be needed to reduce the rate of shunt dysfunction *Ventricular cysts:* Endoscopic resection of cysts. Albendazole may be used only in small lesions located in lateral ventricles. Ventricular shunt only needed in patients with associated ependymitis *Angiitis, chronic arachnoiditis:* No need for cysticidal drug therapy. Corticosteroids are mandatory *Cysticercosis of the spine:* Surgical resection of lesions. Anecdotal use of albendazole with good results

Level 1 of evidence favors the use of cysticidal drugs in patients with parenchymal brain vesicular and colloidal cysts. For other forms of the disease, guidelines are based on Levels 2 and 3 of evidence

Therapy often includes a combination of symptomatic drugs, cysticidal therapy, and surgery (Table 9.1).

9.1 Symptomatic Management

Neurocysticercosis-associated seizures seem to respond very well to a first-line AED (Del Brutto et al. 1992a, b). Carbamazepine and phenytoin are the most frequently used drugs, followed by valproate. When choosing the AED, it must be

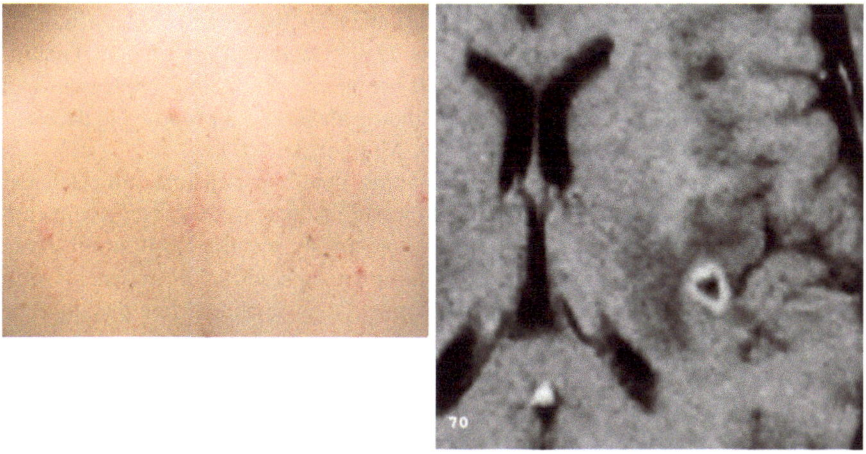

Fig. 9.1 Cutaneous reaction (*left*) in the back of patient with single cysticercus granuloma (*right*) occurring 2 weeks after the start of phenytoin therapy

remembered that some patients with a single cysticercus granuloma may develop transient cutaneous reactions with the use of sodium phenytoin (Fig. 9.1) probably related to mechanisms involved in the immune-mediated response of the host to the parasite (Singh et al. 2004). Irrespective of the AED used, good seizure control is almost always the rule in patients with neurocysticercosis, and refractory seizures are rare. However, there is no consensus on the required length of AED treatment in patients with neurocysticercosis. Some authors have suggested that short-term (3–6 months) use of AEDs—while inflammation around a cyst still exists—may be enough. So far, no controlled evidence exists to support this claim and standard guidelines on the use of AEDs should be followed. After seizure remission and appropriate tapering of AED, still 50 % of patients will present further seizure episodes and require reinstalling AEDs for much longer time (Del Brutto 1994). There is increasing evidence favoring the concept that patients who develop calcifications (either spontaneously or as the result of cysticidal drug therapy) are those that will need longer (even lifelong) courses of AED therapy to reduce the risk of relapses in the long-term follow-up (Verma and Misra 2006).

Headache should be appropriately assessed to rule out intracranial hypertension. If no signs of intracranial hypertension exist, common analgesics should be used to obtain appropriate pain control. The use of anti-inflammatory drugs agents has been mostly restricted to corticosteroids, although early reports on the use of dextrochloropheniramine or, more recently, of methotrexate also exist (Agapejev et al. 1989; Keiser and Nash 2003). Acute anti-inflammatory therapy in neurocysticercosis is primarily directed to reduce edema and local inflammation, thus ameliorating intracranial hypertension and decreasing the propensity to further seizures. Corticosteroids also remain the first line of therapy for patients with cysticercotic angiitis to reduce the risk of arterial occlusion with the subsequent development of a cerebral infarction (Del Brutto 2008). The length of corticosteroid therapy varies according to the size of the lesions and the extent of the resulting inflammatory reaction. In many cases, the use of such drugs should be sustained for weeks and perhaps months (more markedly so in patients with multiple large cysts or extraparenchymal neurocysticercosis). As long-term corticosteroid use is associated with unpleasant and sometimes risky side effects, the use of methotrexate has been proposed to decrease the dose and time of corticosteroid therapy (Mitre et al. 2007).

Particular attention should be given to intracranial hypertension, and cysticidal drugs must not be initiated if intracranial pressure is increased as it may worsen it in the short term (Rangel et al. 1987). As previously mentioned, if severe brain edema is suspected or demonstrated by neuroimaging studies—as in patients with cysticercotic encephalitis—corticosteroids must be given as first-line therapy. The choice of a particular drug and the regimen of therapy are not based on controlled studies, but most experts recommend the use of dexamethasone at doses ranging from 6 to 16 mg/day or even higher doses. Using this approach, a rapid response is usually obtained and most patients improve their condition in a matter of hours or a few days. However, symptom relapses may occur at the time of corticosteroid interruption so gradual tapering of these drugs should be considered. If ventricular cysts or hydrocephalus are present or suspected, or if large cyst or a clump of cysts are present, neurosurgical evaluation is mandatory and shunt placement or resection of the lesion(s) may be required (see below).

9.2 Surgical Management

Before the introduction of modern neuroimaging techniques and cysticidal drugs, patients with suspected neurocysticercosis usually went to the operating room for confirming the diagnosis and to get a definitive cure for the disease (Trelles and Trelles 1978). While over the past 40 years less and less patients with neurocysticercosis underwent resection of lesions, surgery still plays an important role in the management of patients with some specific forms of the disease (Sinha and Sharma 2012).

Hydrocephalus secondary to cysticercotic arachnoiditis always requires the placement of a ventricular shunt. The main complication of ventricular shunt is the high incidence of shunt dysfunction. Actually, the high mortality rate (up to 50 %) associated with the development of hydrocephalus due to neurocysticercosis is related to the number of surgical interventions to revise the shunt (Sotelo and Marin 1987). It is probably that long-term corticosteroid administration may reduce the incidence of shunt dysfunction in these cases (Suastegui-Roman et al. 1996). It has also been suggested that an inversion of CSF transit is the most important cause of shunt dysfunction, as it allows parasitic debris to enter the ventricular system. Therefore, the use of a shunt that functions at a constant flow rate—avoiding the entrance of subarachnoid CSF into the ventricular system—could reduce the number of shunt failures (Sotelo et al. 1995).

Surgery is an option for the resection of ventricular cysticerci, particularly those located within the third and the fourth ventricles. While cysticidal drugs destroy many ventricular cysts, the ensued inflammatory reaction that occur as the result of therapy may cause acute hydrocephalus if those cysts are located near the interventricular foramina of Monro or the cerebral aqueduct. In such cases, ventricular cysts may be more safely removed by surgical excision or endoscopic aspiration (Sinha and Sharma 2012). The surgeon must consider the possibility of cyst migration between the time of diagnosis and the surgical procedure, and this must be ruled out by a neuroimaging study before surgery to avoid unnecessary craniotomies (Zee et al. 1984). Shunt placement should follow or even precede the excision of ventricular cysts associated with ependymitis.

Whether spinal cysticerci (located intramedullary or at the spinal subarachnoid space) should be operated or not is still a matter of debate, and it is likely that this question should never be definitively answered based on the rarity of these forms of the disease. Anecdotal cases suggest that medical treatment may be safe and effective (Garg and Nag 1998). However, most experts prefer to refer those patients to the neurosurgeon until further studies are available. For those patients with spinal subarachnoid cysts, the surgeon must have the same precaution than for ventricular cysts, i.e., to perform neuroimaging studies just before surgery, as those cysts may have been moved from the time of diagnosis (Hernández-Gonzalez et al. 1990).

9.3 Specific (Etiological) Management

9.3.1 Cysticidal Drugs

As will be discussed later on, two drugs (albendazole and praziquantel) have been shown to have potent cysticidal effects in humans (Del Brutto 2003). Other drugs, such as metrifonate and flubendazole have been eventually tried in this condition but their use has been promptly abandoned due to side effects or lack of efficacy (Salazar and Gonzalez 1972; Téllez-Girón et al. 1984). Oxfendazole, a benzimidazole molecule with a long half-life, has been shown to cause a significant clearance of muscle and brain cysts in swine after a single-dose trial (Gonzalez et al. 1997); this drug is now undergoing Phase I studies in humans, and if found safe, it may provide a new therapeutic alternative to albendazole or praziquantel. There is also a single report suggesting that ivermectin may be effective in selected cases (Díazgranados-Sánchez et al. 2008).

Praziquantel is a pyrazino-isoquinoline derivative. It affects calcium channels in the parasite's surface and produces muscle contractions, paralysis, and damage of the tegument (Overbosh et al. 1987). Maximal serum levels are obtained 1.5–2 h after administration and drop rapidly thereafter (Jung et al. 1991). Praziquantel is metabolized in the liver, and size effects are mild and mainly related to gastric disturbances, dizziness, drowsiness, fever, headache, increased sweating, and, less commonly, allergic reactions (reference). For treatment of neurocysticercosis, praziquantel is usually given at daily doses of 50 mg/kg for 2 weeks, with a ceiling of 3 g/day. A single-day course of 75–100 mg/kg divided in three doses given every 2 h has been also described (Corona et al. 1996; Del Brutto et al. 1999), but it seems to be effective in patients with a single lesion but not in those with multiple cysts (Pretell et al. 2001).

Albendazole is a typical benzimidazole compound. It leads to selective degeneration of parasite cytoplasmic microtubules, affecting ATP formation, and also impairs glucose intake, leading to energy depletion and parasite starvation (Rossignol 1981). It binds to tubulin and thus also interferes with parasite cell division. Albendazole is not active by itself and the active molecule, albendazole sulfoxide, results from liver metabolism of albendazole (Castro et al. 2009). Maximal levels are obtained from 2 to 3 h after ingestion, and albendazole penetrates in the CSF better than praziquantel (Jung et al. 1990a). Side effects in humans are mostly related to liver toxicity (increase in liver enzymes), hematological effects, hair loss, and gastrointestinal symptoms (Rossignol 1981). For therapy of neurocysticercosis, albendazole is usually given at daily doses of 15 mg/kg for 1–4 weeks. In the USA, a ceiling of 800 mg/day is frequently used based on FDA-approved doses, while in most other countries, the usual ceiling is 1,200 mg/day (García and Del Brutto 2005).

9.3.2 Trials of Cysticidal Drugs

Cysticidal drugs have been used for therapy of human cysticercosis after Robles and Chavarría (1979) reported the cure of a single Mexican patient with multiple ring-enhancing lesions after a trial with praziquantel. This pioneer report was followed by a number of case series and controlled studies (using historical controls) showing the efficacy of the drug (Botero and Castaño 1982; Sotelo et al. 1984, 1985; Spina-França et al. 1982). Such results prompted clinicians to widely use praziquantel in different forms of the disease. It was then noticed that a few days after the onset of therapy, symptoms increased in a sizable proportion of cases, even with serious adverse effects including intracranial hypertension or death (Wadia et al. 1988). This created confusion among some physicians involved in the care of neurocysticercosis patients, who questioned the benefits of praziquantel and even considered that this drug may actually be deleterious in this setting (Kramer 1995). Meanwhile, albendazole, another cysticidal drug, was first tested in Mexico and then in other disease-endemic countries, and proved effective for destroying parenchymal brain cysticerci, with the advantages of being cheaper and somewhat more effective than praziquantel (Alarcón et al. 1989; Cruz et al. 1991; Escobedo et al. 1987; García et al. 1997; Sotelo et al. 1988, 1990; Takayanagui and Jardim 1992). Some early studies also found a clinical effect on seizure control, with 50–80 % of treated patients being free of seizures compared to 25 % of untreated patients (Del Brutto et al. 1992a, b; Vazquez and Sotelo 1992). However, a couple of controlled studies, published during the 1990s, challenged the usefulness of cysticidal drugs, reporting no significant effect in parasite destruction or in the reduction of seizures (Salinas et al. 1999). Further placebo-controlled studies confirmed the efficacy of cysticidal drugs (as shown in pioneer trials) and settled their actual value in the management of patients with neurocysticercosis (Baranwal et al. 1998; García et al. 2004; Gogia et al. 2003; Kalra et al. 2003).

A review of the published literature including case series of patients with viable parenchymal cysticercosis confirmed by serology and evaluated with neuroimaging studies from 3 to 6 months after the trial shows that praziquantel destroys 57.1 %

(95 % C.I. 55–59 %) while albendazole destroys 71.6 % (95 % C.I. 70–73 %) of lesions. In general terms, 50.5 % (95 % C.I. 45–56 %) of patients treated with praziquantel and 53.1 % (95 % C.I. 48–58 %) of those treated with albendazole were free of lesions after therapy (Hector H. García, unpublished data). Although the above numbers incorporate diverse case series with individual biases and likely overestimate the efficacy of both drugs, these overall estimates agree with most reviews on the subject, suggesting that albendazole has a slightly higher cysticidal efficacy than praziquantel.

In order to better understand the apparent discrepancies in the reported trials, we first need to put them in the context of the different types of neurocysticercosis. The clinical evolution of patients with a single parenchymal brain granuloma is extremely different from that of those with multicystic parenchymal disease or extraparenchymal neurocysticercosis, and the results of trials in different forms of neurocysticercosis cannot meaningfully be pulled together.

9.3.2.1 Cysticidal Drugs for Therapy of Patients with Viable Brain Cysts

As previously noted, the initial trials on cysticidal drug therapy were performed in patients with established brain cysts, many of them with multiple cysts in the vesicular or colloidal stage (neuroimaging studies showing ring-enhancing lesions with a clear hypodense/hypointense center). This type of lesion will not resolve by itself in the short term and is likely to continue causing symptoms for years (García and Del Brutto 2005). It is unclear whether patients with untreated cystic disease will eventually develop subarachnoid disease years later. For clinicians familiar with this form of neurocysticercosis, the benefits of destroying all cysts under controlled conditions are evident, and thus most experts in Latin America—where viable parenchymal brain cysts is a frequent presentation—are inclined in favor of the routine use of cysticidal drugs (Fig. 9.2). Despite the large body of open controlled studies and case series, only a few years ago, the first double-blind, placebo-controlled, randomized trial demonstrated that viable cysts were not expected to resolve by natural evolution in the short term and that albendazole was associated to fewer seizures with generalization in the long term (García et al. 2004). Of note, the efficacy of albendazole in this trial was only 65 %, and the difference in numbers of partial seizures was not statistically significant.

9.3.2.2 Cysticidal Drugs for Therapy of Patients with a Single Cysticercal Granuloma

In the Indian subcontinent, the most frequent form of neurocysticercosis is a single degenerating brain parasite presenting with recent onset seizures in older children and teenagers (García et al. 2010). This type of neurocysticercosis is also seen in infants and children elsewhere in the world (Del Brutto 2013) and also in residents of non-endemic countries who get exposed to the parasite while travelling or due to close contact with an asymptomatic *Taenia solium* carrier living in their close environments (Del Brutto 2012; Del Brutto et al. 2012). Therefore, some Indian experts and pediatricians in non-endemic countries were prone to question the benefits of cysticidal drug therapy based on the demonstrated risks of treatment in a disease

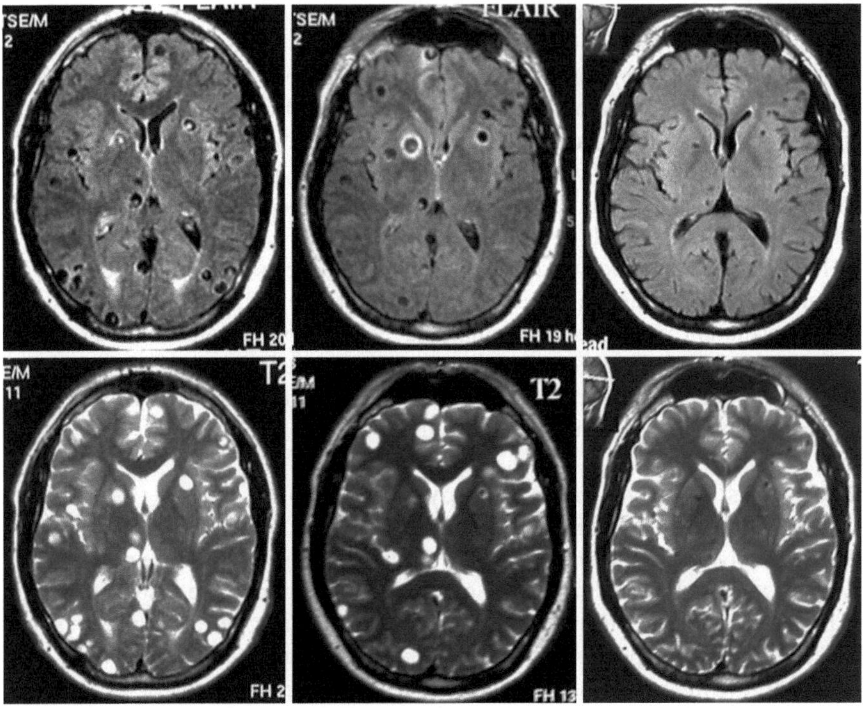

Fig. 9.2 Fluid-attenuated inversion recovery (FLAIR) and T2-weighted MRI showing multiple parenchymal brain cysticerci in the vesicular stage (*left*). Three months after albendazole therapy, most lesions—particularly those located at the occipital lobes—disappeared (*center*), and 1 year after therapy, all lesions have been resolved and many of them have been replaced by a residual calcification (*right*)

which they considered mild and of a very good prognosis (Mitchell and Crawford 1988; Padma et al. 1994). In fact, the risks for seizure relapse in patients with a single cysticercus granuloma seem to be below 30 %, strongly associated to cases where a residual calcification is left instead of total clearance of the brain lesion (Verma and Misra 2006). Given the already-established inflammatory reaction with cellular infiltrate and cyst destruction and the fact that most of these lesions frequently resolve without specific treatment, it is reasonable to conclude that cysticidal drugs would have minimal or no effect in this form of neurocysticercosis (Singh et al. 2010). The problem has been mainly generated by the incorrect recognition of a single cysticercus granuloma and by the fact that some studies have probably included patients with both colloidal and granular cysticerci.

9.3.2.3 Cysticidal Drugs for Therapy of Patients with Extraparenchymal Lesions

While some studies have shown that albendazole may successfully destroy ventricular cysticerci (Del Brutto and Sotelo 1990; Proaño et al. 1997), endoscopic resection of the lesion seems to be the safest and most effective approach for this form of the disease (Sinha and Sharma 2012). The doubt remains on whether patients should receive a trial with cysticidal drugs after resection of ventricular cysts (Khade et al. 2013). If ventricular cysts were entirely removed, there should be no reason for further medical treatment. However, there is no certainty that other smaller lesions are missed by neuroimaging studies, particularly in the ventricles or basal CSF cisterns. Persistence of cysts or membrane remnants could lead to further clinical issues and later disease manifestations. Use of circulating antigen detection assays could orientate the use of cysticidal drug therapy in such patients. Alternatively, a trial with cysticidal drugs could be proposed on the basis that if there are no remaining parasites, the risks associated with treatment are minimal or nonexistent. In any case, neuroimaging and serological surveillance of these patients should be kept for a few years.

Patients with involvement of CSF cisterns at the base of the brain by cysticerci usually develop protracted courses and grim prognoses. In these cases, after hydrocephalus and intracranial hypertension have been managed, cysticidal drugs should be given until no evidence of living parasites exists anymore. Published evidence suggests that albendazole is the preferred drug in this setting (Del Brutto 1997; Del Brutto et al. 1992b), although some patients have not responded to medical treatment (Cardenas et al. 2010). Treatment of giant cysts located inside CSF cisterns usually requires higher doses of albendazole (up to 30 mg/kg/day), more prolonged courses of therapy (more than 1 month), or even repeated cycles (Góngora-Rivera et al. 2006; Proaño et al. 2001). Again, neuroimaging and circulating antigen monitoring would be important here since antibodies will persist for years. Caution should be taken in these patients because treatment may result in sudden intracranial hypertension or stroke (Noboa 1993). However, under appropriate corticosteroid coverage, these events are rare and the risk of disease progression outweighs those of treatment.

9.3.3 Meta-Analysis Data on Cysticidal Drug Therapy for Neurocysticercosis

Six meta-analyses have been published to date. The first of them included only four randomized trials comparing albendazole or praziquantel versus placebo or no treatment in patients with parenchymal brain cysts; the authors found no differences in the frequency of seizures and a lower risk of persistence of cysts among treated versus nontreated patients (Salinas et al. 1999). The second meta-analysis evaluated 11 randomized trials comparing the same outcomes than in the former, but here, the use of cysticidal drugs were associated with a better resolution of colloidal and vesicular cysticerci, a lower risk for recurrence of seizures in patients with colloidal cysticerci, and a reduction in the rate of generalized seizures in those with vesicular cysticerci (Del Brutto et al. 2006). The third meta-analysis included four randomized trials and ten observational studies for evaluating the efficacy of albendazole in children with parenchymal brain enhancing lesions, showing reduction in the relative risk of seizure recurrence in treated patients (Mazumdar et al. 2007). A fourth meta-analysis compared the efficacy of albendazole versus that of praziquantel in six trials and found that albendazole is more effective than praziquantel for reducing the number of recurrent seizures, but the efficacy of both drugs for destroying cystic lesions is rather similar, with some trials favoring albendazole while others favoring praziquantel (Matthaiou et al. 2008). The fifth meta-analysis—published as part of the Cochrane library—showed that the use of albendazole reduces the number of cystic lesions and the chance of seizure recurrence; this meta-analysis combined series of patients with viable cysts and with enhancing lesion, so the compound result is difficult to interpret (Abba et al. 2010). Finally, the sixth meta-analysis reviewed 15 randomized controlled trials comparing rates of seizure freedom and single-granuloma resolution in patients receiving cysticidal drugs or corticosteroids versus those that were only treated with AEDs, and it was found that the use of albendazole improved the rate of seizure-free persons and hastens the resolution of the granuloma when compared with no treatment and that the role of corticosteroids alone remains uncertain (Otte et al. 2013).

Recently, a group of US neurologists revised published data to issue a guideline document for the American Academy of Neurology, concluding that albendazole plus corticosteroids should be considered for patients (adults and children) with neurocysticercosis, as the use of these drugs reduce the number of viable cysts on control neuroimaging studies (level B of evidence) and the long-term risk of seizure recurrence (level B evidence). According to the panel, available data is insufficient to support or restrict the use of corticosteroids alone for therapy of patients with parenchymal brain cysticercosis (Baird et al. 2013).

9.3.4 Concomitant Medication During Antiparasitic Treatment

Cysticidal drug treatment may be associated with exacerbation of neurological symptoms due to the damage of the parasite and the breakage of its immune-modulatory mechanisms, with an associated local inflammatory response and perilesional edema. This is particularly true for patients with heavy infections with parenchymal brain cysts and for those with subarachnoid or ventricular neurocysticercosis (Del Brutto et al. 1992b; García and Del Brutto 1999; Proaño et al. 2001). For this reason, corticosteroids are usually coadministered during a trial with cysticidal drugs. Most authors use corticosteroids from the start of the trial or just 1 day before, while others prefer to start these drugs a few days in advance (Márquez-Caraveo et al. 2004). Corticosteroids are often given at moderate doses (most commonly 6–12 mg/day of dexamethasone or 20 mg of methylprednisolone per day) and tapered after the end of cysticidal drug therapy. Since corticosteroids cause gastric irritation, omeprazole at doses of 20 mg/day is often given to minimize this side effect. Therapy with AEDs should be optimized and maintained as *per standard* use. Plasma levels of albendazole or praziquantel are extremely variable from one individual to another and seem to depend on several factors, including the simultaneous use of corticosteroids (Jung et al. 1990b; Takayanagui et al. 1997; Vazquez et al. 1987).

Some authors propose the use of fatty meals to increase plasma levels of albendazole sulfoxide, or grapefruit juice to increase praziquantel levels (Lange et al. 1988; Castro et al. 2002).

9.3.5 Treatment of Asymptomatic Neurocysticercosis

There is absolutely no controlled data to base treatment decisions on the best therapeutic approach to asymptomatic individuals diagnosed of viable neurocysticercosis when neuroimaging studies are performed by an unrelated reason, i.e., head trauma or detected during the screening of *Taenia solium* carriers' household contacts. The risks and benefits of any specific treatment should then be weighted and considered in the context of the individual infection type and parasite burden. Cysticidal drugs can trigger seizures or—in patients with massive infections or huge intracranial cysts—lead to intracranial hypertension, hydrocephalus, or stroke. On the other hand, as much as it is known, extraparenchymal neurocysticercosis is progressive and eventually lethal, and no alternative therapies exist, so the use of cysticidal drugs in these patients, even at an early, asymptomatic stage, should be seriously considered. If a decision not to treat is done, neuroimaging and immunological surveillance should be enforced. A more conservative approach can be considered in patients with a single and small parenchymal brain cyst.

9.3.6 Further Directions of Cysticidal Treatment

There are preliminary reports on the use of the combination of albendazole plus praziquantel for the treatment of neurocysticercosis. In the initial report, Guo et al. (2003) found markedly higher cysticidal efficacy in the combined regimen. A further study in Indian children with single enhancing lesions also found a nonsignificant trend for a better efficacy when both drugs were given (Kaur et al. 2009). The superior efficacy of this combination is biologically plausible since both drugs have different mechanisms of action and their combined use may provide better clearance of cysticerci (García et al. 2011). An ongoing, double-blind, large-scale study in patients with viable parenchymal neurocysticercosis will help to resolve the question of whether two drugs are better than one for treatment of this form of the disease (Hector H. García, unpublished results).

References

Abba K, Ramaratnam S, Ranganathan LN (2010) Anthelminthics for people with neurocysticercosis. Cochrane Database Syst Rev (1):CD000215.

Agapejev S, Meira DA, Barraviera B, Machado JM, Marques PC, Mendes RP, Kamegasawa A, Ueda AK (1989) Neurocysticercosis: treatment with albendazole and dextrochloropheniramine. Trans R Soc Trop Med Hyg 83:377–383

Alarcón F, Escalante L, Dueñas G, Montalvo M, Román M (1989) Neurocysticercosis. Short course of treatment with albendazole. Arch Neurol 46:1231–1236

Baird RA, Zunt JR, Halperin JJ, Gronseth G, Roos KL (2013) Evidence-based guideline: treatment of parenchymal neurocysticercosis. Report of the guideline development subcommittee of the American Academy of Neurology. Neurology 80:1424–1429

Baranwal AK, Singhi PD, Khandelwal N, Singhi SC (1998) Albendazole therapy in children with focal seizures and single small enhancing computerized tomographic lesions: a randomized, placebo-controlled double-blind trial. Pediatr Infect Dis 17:696–700

Botero D, Castaño S (1982) Treatment of cysticercosis with praziquantel in Colombia. Am J Trop Med Hyg 31:811–821

Cardenas G, Carrillo-Mezo R, Jung H, Sciutto E, Soto-Hernandez JL, Fleury A (2010) Subarachnoid neurocysticercosis non-responsive to cysticidal drugs: a case series. BMC Neurol 10:16

Castro N, Jung H, Medina R, Gónzalez-Esquivel D, Lopez M, Sotelo J (2002) Interaction between grapefruit and praziquantel in humans. Antimicrob Agents Chemother 46:1614–1616

Castro N, Márquez-Caraveo C, Brundage RC, González-Esquivel D, Suárez AM, Góngora F, Jara A, Urizar J, Lanao JM, Jung H (2009) Population pharmacokinetics of albendazole in patients with neurocysticercosis. Int J Clin Pharmacol Ther 47:679–685

Corona T, Lugo R, Medina R, Sotelo J (1996) Single-day praziquantel therapy for neurocysticercosis. N Engl J Med 334:125

Cruz M, Cruz I, Horton J (1991) Albendazole vs praziquantel in the treatment of cerebral cysticercosis: clinical evaluation. Trans R Soc Trop Med Hyg 85:244–247

Del Brutto OH (1994) Prognostic factors for seizure recurrence after withdrawal of antiepileptic drugs in patients with neurocysticercosis. Neurology 44:1706–1709

Del Brutto OH (1997) Albendazole therapy for subarachnoid cysticerci: clinical and neuroimaging analysis of 17 patients. J Neurol Neurosurg Psychiatry 62:659–661

Del Brutto OH (2003) Medical management of neurocysticercosis. Int J Antimicrob Ag 3:133–137

Del Brutto OH (2008) Stroke and vasculitis in patients with cysticercosis. In: Caplan LR (ed) Uncommon causes of stroke. Cambridge University Press, New York

Del Brutto OH (2012) Neurocysticercosis among international travelers to disease-endemic areas. J Travel Med 19:112–117

Del Brutto OH (2013) Neurocysticercosis in infants and toddlers. Report of seven cases and review of published patients. Pediatr Neurol 48(6):432–435

Del Brutto OH, Sotelo J (1990) Albendazole therapy for subarachnoid and ventricular cysticercosis. J Neurosurg 72:816–817

Del Brutto OH, Santibáñez R, Noboa CA, Aguirre R, Díaz E, Alarcón TA (1992a) Epilepsy due to neurocysticercosis: analysis of 203 patients. Neurology 42:389–392

Del Brutto OH, Sotelo J, Aguirre R, Díaz-Calderón E, Alarcón TA (1992b) Albendazole therapy for giant subarachnoid cysticerci. Arch Neurol 49:535–538

Del Brutto OH, Campos X, Sánchez J, Mosquera A (1999) Single-day praziquantel versus 1-week albendazole for neurocysticercosis. Neurology 52:1079–1081

Del Brutto OH, Roos KL, Coffey CS, García HH (2006) Meta-analysis: cysticidal drugs for neurocysticercosis: albendazole and praziquantel. Ann Intern Med 145:43–51

Del Brutto OH, Nash TE, García HH (2012) Cysticerci-related single parenchymal brain enhancing lesions in non-endemic countries. J Neurol Sci 319:32–36

Díazgranados-Sánchez JA, Barrios-Arrázola G, Costa JL, Burbano-Pabon J, Pinzón-Bedoya J (2008) Ivermectina como alternativa terapéutica en neurocysticercosis resistente al tratamiento farmacológico convencional. Rev Neurol 46:671–674

Escobedo F, Penagos P, Rodriguez-Carbajal J, Sotelo J (1987) Albendazole therapy for neurocysticercosis. Arch Intern Med 147:738–741

Estañol B, Kleriga E, Loyo M, Mateos H, Lombardo L, Gordon F, Saguchi AF (1983) Mechanisms of hydrocephalus in cerebral cysticercosis: implications for therapy. Neurosurgery 13: 119–123

García HH, Del Brutto OH (1999) Heavy nonencephalitic cerebral cysticercosis in tapeworm carriers. Neurology 53:1582

García HH, Del Brutto OK (2005) Neurocysticercosis: updated concepts about an old diseases. Lancet Neurol 4:653–661

García HH, Gilman RH, Horton J, Martinez M, Herrera G, Altamirano J, Cuba JM, Rios-Saavedra N, Verastegui M, Boero J, Gonzalez AE (1997) Albendazole therapy for neurocysticercosis: a prospective double-blind trial comparing 7 versus 14 days of treatment. Neurology 48:1421–1427

García HH, Evans CAW, Nash TE, Takayanagui OM, White CA Jr, Botero D, Rajshekhar V, Tsang VCW, Schantz PM, Allan JC, Flisser A, Correa D, Sarti E, Friedland JS, Martinez SM, Gonzalez AE, Gilman RH, Del Brutto OH (2002) Current consensus guidelines for treatment of neurocysticercosis. Clin Microbiol Rev 15:747–756

García HH, Pretell EJ, Gilman RH, Martinez SM, Moulton LH, Del Brutto OH, Herrera G, Evanas CWA, Gonzalez AE (2004) A trial of antiparasitic treatment to reduce the rate of seizures due to cerebral cysticercosis. N Engl J Med 350:249–258

García HH, Gonzalez AE, Rodriguez S, Tsang VC, Pretell EJ, Gonzales I, Gilman RH (2010) Neurocysticercosis: unraveling the nature of the single cysticercal granuloma. Neurology 75:654–658

García HH, Lescano AG, Lanchote VL, Pretell EJ, Gonzales I, Bustos JA, Takayanagui OM, Bonato PS, Horton J, Saavedra H, Gonzalez AE, Gilman RH (2011) Pharmacokinetics of combined treatment with praziquantel and albendazole in neurocysticercosis. Br J Clin Pharmacol 72:77–84

Garg RK, Nag D (1998) Intramedullary spinal cysticercosis: response to albendazole: case reports and review of literature. Spinal Cord 36:67–70

Gogia S, Talkdar B, Choudhury V, Arora BS (2003) Neurocysticercosis in children: clinical findings and response to albendazole therapy in a randomized, double-blind, placebo-controlled trial in newly diagnosed cases. Trans R Soc Trop Med Hyg 97:416–421

Góngora-Rivera F, Soto-Hernández JL, Gonzalez-Esquivel D et al (2006) Albendazole trial at 15 or 30 mg/kg/day for subarachnoid and intraventricular cysticercosis. Neurology 66:436–438

Gonzalez AE, Falcon N, Gavidia C, García HH, Tsang VC, Bernal T, Romero M, Gilman RH (1997) Treatment of porcine cysticercosis with oxfendazole: a dose-response trial. Vet Rec 141:420–422

Guo Y, Xie SP, Jia JP (2003) Therapeutic efficacy of praziquantel, albendazole and a combination of the two drugs in cysticercosis (article in Chinese). Zhongguo Ji Sheng Chong Xue Ji Sheng Chong Bing Za Zhi 21:187–188

Hernández-Gonzalez LA, Arredondo-Mendoza F, Prado-Castro JA (1990) Neurocisticercosis raquimedular en Guatemala. Descripción del signo de "lesión quística flotante". Rev Mex Radiol 44:165–169

Jung H, Hurtado M, Medina MT, Sanchez M, Sotelo J (1990a) Plasma and CSF levels of albendazole and praziquantel in patients with neurocysticercosis. Clin Neuropharmacol 13:559–564

Jung H, Hurtado M, Medina MT, Sanchez M, Sotelo J (1990b) Dexamethasone increases plasma levels of albendazole. J Neurol 237:279–280

Jung H, Vazquez ML, Sanchez M, Penagos P, Sotelo J (1991) Clinical pharmacokinetics of praziquantel. Proc West Pharmacol Soc 34:335–340

Kalra V, Dua T, Kumar V (2003) Efficacy of albendazole and short-course dexamethasone treatment in children with 1 or 2 ring-enhancing lesions of neurocysticercosis: a randomized controlled trial. J Pediatr 143:111–114

Kaur S, Singhi P, Singhi S, Khandelwal N (2009) Combination therapy with albendazole and praziquantel versus albendazole alone in children with seizures and single lesion neurocysticercosis: a randomized, placebo-controlled double blind trial. Pediatr Infect Dis J 28:403–406

Keiser PB, Nash TE (2003) Prolonged perilesional edema after treatment of parenchymal neurocysticercosis: methotrexate as a corticosteroid-sparing agent. Clin Infect Dis 36:e122–e126

Khade P, Lemos RS, Toussaint LG (2013) What is the utility of postoperative antihelminthic therapy after resection of intraventricular neurocysticercosis? World Neurosurg 79:558–567

Kramer D (1995) Medical treatment of cysticercosis – ineffective. Arch Neurol 52:101–102

Lange H, Eggers R, Bircher J (1988) Increased systemic availability of albendazole when taken with a fatty meal. Eur J Clin Pharmacol 34:315–317

Márquez-Caraveo C, Góngora-Rivera F, Santos Zambrano J, Hernández R, Soto-Hernández JL (2004) Pre-treatment with corticosteroids and a single cycle of high dose albendazole for subarachnoidal cysticercosis. J Neurol Neurosurg Psychiatry 75:938–939

Matthaiou DK, Panos G, Adamidi ES, Falagas ME (2008) Albendazole versus praziquantel in the treatment of neurocysticercosis: a meta-analysis of comparative trials. PLOS Negl Trop Dis 2(3):e194

Mazumdar M, Pandharipande P, Poduri A (2007) Does albendazole affect seizure remission and computed tomography response in children with neurocysticercosis? A systematic review and meta-analysis. J Child Neurol 22:135–142

Mitchell WG, Crawford TO (1988) Intraparenchymal cerebral cysticercosis in children: diagnosis and treatment. Pediatrics 82:76–82

Mitre E, Talaat KR, Sperling MR, Nash TE (2007) Methotrexate as a corticosteroid-sparing in complicated neurocysticercosis. Clin Infect Dis 44:549–553

Nash TE, García HH (2011) Diagnosis and treatment of neurocysticercosis. Nat Rev Neurol 7:584–594

Nash TE, Pretell EJ, Lescano AG, Bustos JA, Gilman RH, Gonzalez AE, García HH (1999) Perilesional brain oedema and seizure activity in patients with calcified neurocysticercosis: a prospective cohort and nested case-control study. Lancet Neurol 7:1099–1105

Nash TE, Del Brutto OH, Butman JA, Corona T, Delgado-Escueta A, Duron RM, Evans CAW, Gilman RH, Gonzalez AE, Loeb JA, Medina MT, Pietsch-Escueta S, Pretell EJ, Takayanagui OM, Theodore W, Tsang VCW, García HH (2006a) Calcific neurocysticercosis and epileptogenesis. Neurology 62:1934–1938

Nash TE, Singh G, White AC Jr, Rajshekhar V, Loeb JA, Proaño JV, Takayanagui OM, Gonzalez AE, Butman JA, DeGiorgio C, Del Brutto OH, Delgado-Escueta A, Evans CAW, Gilman RH, Martinez SM, Medina MT, Pretell EJ, Teale J, García HH (2006b) Treatment of neurocysticercosis. Current status and future research needs. Neurology 67:1120–1127

Noboa C (1993) Albendazole therapy for giant subarachnoid cysticerci. Arch Neurol 50: 347–348

Otte WM, Singla M, Sander JW, Singh G (2013) Durg therapy for solitary cysticercus granuloma: a systematic review and meta-analysis. Neurology 80:152–162

Overbosh D, van de Nes JCM, Groll E, Diekmann HW, Polderman AM, Mattie H (1987) Penetration of praziquantel into cerebrospinal fluid and cysticerci in human cysticercosis. Eur J Clin Pharmacol 33:287–292

Padma MV, Behari M, Misra NK, Ahuja GK (1994) Albendazole in single CT ring lesions in epilepsy. Neurology 44:1344–1346

Pretell EJ, García HH, Gilman RH, Saavedra H, Martínez M (2001) Failure of one-day praziquantel treatment in patients with multiple neurocysticercosis lesions. Clin Neurol Neurosurg 103:175–177

Proaño JV, Madrazo I, Garcia L, Garcia-Torres E, Correa D (1997) Albendazole and praziquantel treatment in neurocysticercosis of the fourth ventricle. J Neurosurg 87:29–33

Proaño JV, Madrazo I, Avelar F, López-Félix B, Díaz G, Grijalva I (2001) Medical treatment for neurocysticercosis characterized by giant subarachnoid cysts. N Engl J Med 345:879–885

Rangel R, Torres B, Del Bruto O, Sotelo J (1987) Cysticercotic encephalitis: a severe form in young females. Am J Trop Med Hyg 36:387–392

Robles C, Chavarría M (1979) Presentación de un caso clínico de cisticercosis cerebral curado médicamente con un nuevo fármaco: praziquantel. Salud Pub Mex 21:603–618

Rossignol JF (1981) Albendazole: estudios clínicos realizados en Francia y Africa Occidental. Informe sobre 1034 casos. Compendium de Investigaciones Clínicas Latinoamericanas 1:117–125

Salazar M, Gonzalez D (1972) Ensayo terapéutico de la cisticercosis cerebral con metrifonato. Rev Invest Salud Pub (Mex) 32:1–7

Salinas R, Counsell C, Prasad K, Gelband H, Garner P (1999) Treating neurocysticercosis medically: a systematic review of randomized, controlled trials. Trop Med Int Health 4:713–718

Singh G, Kaushal S, Gupta M, Chander Chopra S (2004) Cutaneous reactions in patients with solitary cysticercus granuloma on phenytoin sodium. J Neurol Neurosurg Psychiatry 75:331–333

Singh G, Rajshekhar V, Murthy JM, Prabhakar S, Modi M, Khandelwal N, García HH (2010) A diagnostic and therapeutic scheme for a solitary cysticercus granuloma. Neurology 75:2236–2245

Sinha S, Sharma BS (2012) Intraventricular neurocysticercosis: a review of current status and management issues. Br J Neurosurg 26:305–309

Sotelo J, Marin C (1987) Hydrocephalus secondary to cysticercotic arachnoiditis. A long-term follow-up of 92 cases. J Neurosurg 66:686–689

Sotelo J, Escobedo F, Rodriguez-Carbajal J, Torres B, Rubio-Donnadieu F (1984) Therapy of parenchymal brain cysticercosis with praziquantel. N Engl J Med 310:1001–1007

Sotelo J, Torres B, Rubio-Donnadieu F, Escobedo F, Rodriguez-Carbajal J (1985) Praziquantel in the treatment of neurocysticercosis: long-term follow-up. Neurology 35:752–754

Sotelo J, Penagos P, Escobedo F, Del Brutto OH (1988) Short course of albendazole therapy for neurocysticercosis. Arch Neurol 45:1130–1133

Sotelo J, Del Brutto OH, Penagos P, Escobedo F, Torres B, Rodríguez-Carabajal J, Rubio-Donnadieu F (1990) Comparison of therapeutic regimen of anticysticercal drugs for parenchymal brain cysticercosis. J Neurol 237:69–72

Sotelo J, Rubalcaba MA, Gomez-Llata S (1995) A new shunt for Hydrocephalus that relies on CSF production rather than on ventricular pressure: initial clinical experience. Surg Neurol 43:324–332

Spina-França A, Nobrega JPS, Livramento JA, Machado LR (1982) Administration of praziquantel in neurocysticercosis. Tropenmed Parasitol 33:1–4

Suastegui-Roman RA, Soto-Hernández JL, Sotelo J (1996) Effects of prednisone on ventriculo-peritoneal shunt function in hydrocephalus secondary to cysticercosis: a preliminary study. J Neurosurg 84:629–633

Takayanagui OM, Jardim E (1992) Therapy for neurocysticercosis. Comparison between albendazole and praziquantel. Arch Neurol 49:290–294

Takayanagui OM, Lanchote VL, Marques MP, Bonato PS (1997) Therapy for neurocysticercosis: pharmacokinetic interactions of albendazole sulphoxide with dexamethasone. Ther Drug Monit 19:51–55

Téllez-Girón E, Ramos MC, Dufour L, Montante M, Tellez E, Rodriguez J, Gómez-Méndez F, Mireles E (1984) Treatment of neurocysticercosis with flubendazole. Am J Trop Med Hyg 33:627–631

Trelles JO, Trelles L (1978) Parasitic diseases and tropical neurology. In: Vinken PJ, Bruyn GW (eds) Handbook of clinical neurology, vol 35. North Holland, Amsterdam

Vazquez V, Sotelo J (1992) The course of seizures after treatment for cerebral cysticercosis. N Engl J Med 327:696–701

Vazquez ML, Jung H, Sotelo J (1987) Plasma levels of praziquantel decrease when dexamethasone is given simultaneously. Neurology 37:1561–1562

Verma A, Misra S (2006) Outcome of short-term antiepileptic treatment in patients with solitary cerebral cysticercus granuloma. Acta Neurol Scand 113:174–177

Wadia N, Desai S, Bhatt M (1988) Disseminated cysticercosis. New observations, including CT scan findings and experience with treatment by praziquantel. Brain 111:597–614

Zee C-S, Segall HD, Apuzzo MLJ, Ahmadi J, Dobkin WR (1984) Intraventricular cysticercal cysts: further neuroradiological observations and neurosurgical implications. ANJR Am J Neuroradiol 5:727–730

Control and Perspectives for Elimination of *Taenia solium* Taeniasis/Cysticercosis 10

As previously discussed in this book, *Taenia solium* is endemic in vast parts of the world, in particular, in regions where pigs are raised as domestic animals in rural areas. The resulting disease, neurocysticercosis, is also present in non-endemic regions because of travel and immigration. Taeniasis/cysticercosis exerts a significant economic impact worldwide, most of which is associated with human neurocysticercosis, although there are other associated costs related to porcine cysticercosis (Bhattarai et al. 2012; Praet et al. 2009; Torgerson and Macpherson 2011). The substantial contribution of neurocysticercosis to the burden of neurological disease in Latin America has been estimated as between 23,512 and 39,186 symptomatic neurocysticercosis cases in Peru, and approximately, 400,000 in Latin America (Bern et al. 1999). Even in non-endemic regions, immigrants with neurocysticercosis may become a significant economic burden for health systems. General costs related to neurocysticercosis in California, USA, have been estimated to be $17 million, with hospital costs above $5 million (Croker et al. 2012).

In disease-endemic regions, neurocysticercosis has been blamed to account for approximately one third of all epilepsies (Del Brutto et al. 2005; Montano et al. 2005; Ndimubanzi et al. 2010) after accounting for the baseline prevalence of serological markers and CT abnormalities in asymptomatic individuals. This proportion is quite consistent among studies (Ndimubanzi et al. 2010). From all causes of epilepsy, only a few are amenable to strategies for their decrease in the medium to long term, and even fewer could be eliminated. At the population level, the numbers of seizure cases related to motor vehicle accidents respond to enforcing legislation, perinatal epilepsies respond to improved prenatal control and better delivery conditions, stroke-associated seizures respond to secondary prevention (Newton and García 2012). None of these conditions can be foreseen to disappear. Conversely, if transmission of *Taenia solium* is eliminated, up to 30 % of all epilepsy cases could be prevented.

Being such a significant public health problem, actions on control and potential elimination of *Taenia solium* are urgently required. Actively intervening to control *Taenia solium* is not a new idea. Control interventions have been tested since 1985, and the International Task Force for Disease Eradication listed *Taenia solium*

O.H. Del Brutto, H.H. García, *Cysticercosis of the Human Nervous System*,
DOI 10.1007/978-3-642-39022-7_10, © Springer-Verlag Berlin Heidelberg 2014

taeniasis/cysticercosis as potentially eradicable (International Task Force for Disease Eradication 1993), mainly because of the lack of contributing wild reservoirs, pig as an easily targetable intermediate host, and the availability of effective treatments for taeniasis (and now for pig cysticercosis), among other reasons. With this, taeniasis cysticercosis joined a reduced number of diseases targeted for elimination and potential eradication.

10.1 Principles of Control

10.1.1 Treatment of Taeniasis

The only source of infection for humans and pigs is the human tapeworm carrier. Moreover, the adult *Taenia solium* tapeworm has a very high biotic potential, meaning that a single worm can infect many hosts in its surroundings. For all these reasons the tapeworm carrier is the main target of control interventions. Tapeworm carriers may be treated and cured either by case diagnosis and individualized treatment or by mass treatment of the human population with a single oral dose of either niclosamide (2 g in adults) or praziquantel (5–10 mg/kg). Diagnosis of taeniasis is usually made by stool microscopy after concentration by sedimentation, with poor sensitivity (under 70 %) (García et al. 2007a). Detection of tapeworm antigens in stools (coproantigen detection), described by Allan et al. (1990), is a much more sensitive technique but is not widely available beyond research settings.

Mass treatment is logistically easier and likely much cheaper. However, there is a theoretical risk that the administration of antihelminthic drugs to an individual who has silent neurocysticercosis could trigger seizures or other neurological symptoms. This has been reported in a few individuals after a single dose of praziquantel (Flisser et al. 1993; Johnson 1986; Torres et al. 1988) or after a single dose of the commonly used antihelminthic albendazole (García et al. 2007b; Ramos-Zuniga et al. 2013). This seems to be a rare event since praziquantel is routinely administered in mass-treatment campaigns for schistosomiasis in Africa, some of which are also endemic for cysticercosis, and reports of neurological side effects are rare (Johnson 1986). Niclosamide does not have this drawback but it is poorly available in many parts of the drug (as it also happens with praziquantel). Adding mass niclosamide or praziquantel treatment to existing mass deworming programs has been suggested as a highly cost-effective control strategy (Alexander et al. 2011).

A series of experiences with mass chemotherapy have been performed in several countries. Between 1985 and 1987, Cruz et al. (1989) de-parasitized over 10,000 individuals in Loja, Ecuador, using a single dose of 5 mg/kg of praziquantel. They found 1.6 % of taeniasis by patient's self-report and published a reduction in porcine cysticercosis after the program. This program, however, was interrupted soon after its implementation, and a few years after stopping control, transmission returned to levels close to baseline. Díaz Camacho et al. (1991)

performed mass praziquantel treatment in a village of 559 individuals in Sinaloa, Mexico. The authors reported no taeniasis cases or cysticercosis-infected pigs 1 year after the intervention and noted a marked clustering around tapeworm carriers. However, 1 year after the intervention, porcine cysticercosis detected by tongue examination had raised from 6 to 11 % (Keilbach et al. 1989). From 1991 to 1996, Sarti et al. (2000) worked in three villages in Morelos, Mexico. In one of them, mass chemotherapy with 5 mg/kg praziquantel was 67 % effective to treat human taeniasis. The prevalence of taeniasis dropped by half, and there was also a long-term drop in porcine seroprevalence and a decrease in late-onset seizures. Health education was tested in one of these villages and shown to decrease transmission of taeniasis/cysticercosis very fast after the intervention (Sarti et al. 1997). In the third community, where both health education and mass praziquantel treatment were administered together, the prevalence of human taeniasis did not show an impact along time, but porcine prevalence decreased gradually (Pawlowski et al. 2005).

Between 1994 and 1996, Allan et al. (1997) used mass niclosamide therapy in approximately 2,000 people in two villages in Guatemala and dropped the prevalence of taeniasis from 3.5 to 1 %, accompanied by a marked drop in pig seroprevalence. More recently, an educational intervention was also tested in 42 villages (827 households) in Mbulu district, Tanzania. It resulted in improved knowledge of the disease without apparent changes in observed risk practices; however, the incidence of porcine cysticercosis was significantly reduced in the intervention arm (Ngowi et al. 2008). Wide efforts in China using health education, mass chemotherapy, and environment management are reported to have reduced porcine cysticercosis by 97 % and human taeniasis by 91 % along 6 years (Wu et al. 2012).

10.1.2 Treatment of Infected Pigs

Treatment of the human tapeworm carrier ignores the porcine reservoir. Mathematical modeling strongly suggests that driving disease transmission to an extinction state would require multiple sustained interventions in the human population, for a very long time (Gonzalez 1997). Treatment of infected pigs would avoid future taeniasis infections and increase the likelihood of elimination. Flisser and coworkers (1990a, b) demonstrated the efficacy of praziquantel to eliminate muscle cysts in pigs. Tellez-Giron et al. (1981) reported 100 % efficacy with 10 days of flubendazole, Gonzalez et al. (1995) reported good efficacy of multiple doses of albendazole, and Peniche-Cardenas et al. (2002) reported similar effects using 8 days of intramuscular albendazole sulfoxide. Most of these regimes require multiple doses which is highly impractical in field conditions. Oxfendazole, given as a single oral dose of 30 mg/kg, is close to 100 % effective to destroy muscle cysts; it is not expensive and has no major side effects. Death of cysts usually occurs along a few weeks. The withdrawal period of oxfendazole at this dose was calculated to be 17 days. Oxfendazole treatment of pigs at 4 and 9 months of age as a single control measure was tested in two groups of 54 pigs each and 108 controls in Mozambique; significant decreases in pig infection were found, more marked when pigs were treated at 9 months of age (Pondja et al. 2012).

Combined human and pig treatment was applied by the Cysticercosis Working Group in Peru in ten villages in Huancayo, at the Peruvian central highlands. The effect was measured in pig seroincidence and demonstrated a significant decrease followed by a rapid return to levels close to baseline, quick after the intervention was stopped (García et al. 2006).

10.1.3 Pig Vaccination

A third major pathway of intervention is the immunization of the porcine population. Molinari et al. (1993) performed a large trial of pig immunization using a crude cysticercal extract in 3,295 animals, a proportion of which were vaccinated twice. No infected animals were reported 1 year after the intervention, although it is unclear how infection status of pigs was determined. Others have demonstrated diverse degrees of protection using antigens from the scolex, the entire cyst, or from the oncosphere (Nascimento et al. 1995; Plancarte et al. 1999; Verastegui et al. 2002). Improved pig vaccines were later developed. TSOL18 is an oncospheral recombinant protein identified and developed in Australia. Two doses of TSOL18 vaccine provide almost absolute protection to further infection (Flisser et al. 2004; Gonzalez et al. 2005), a protection that was later confirmed in field trials (Assana et al. 2010). A vaccine using synthetic peptides of *Taenia crassiceps* antigens has been produced and applied in Mexico. This vaccine seems to obtain reductions of up to 98.7 % in the numbers of parasites but only 53 % in the proportion of infected pigs (Huerta et al. 2002). DNA vaccines have also been shown to provide diverse degrees of protection, up to 93 % (Guo et al. 2004, 2007; Wang et al. 2003).

10.1.4 Other Potential Points of Intervention

A time-honored form to detect infected pigs in the field is to examine their tongues to see cysticerci nodules (Fig. 10.1). However, not all infected animals have a positive tongue examination. Inspection of carcasses at slaughterhouses is also a classic standard in control which, nevertheless, has limited value to detect porcine cysticercosis in endemic regions. First of all, most cysticercosis infected pigs will have less than five or ten cysts in the entire carcass, so opening the carcass and performing a few cuts in the larger muscles is likely insufficient to detect these few cysts. Furthermore, peasants will be reluctant to bring animals at risk of being infected with cysticercosis because of wrong confiscation policies with no reimbursement to the affected peasant (Cysticercosis Working Group in Peru 1993; Gemmell et al. 1983).

As noted, a major problem in rural villages of developing countries is the fact that pigs are allowed to roam free and the close relationship that they have with humans (Fig. 10.2). So, pigs get access to human feces in places with poor

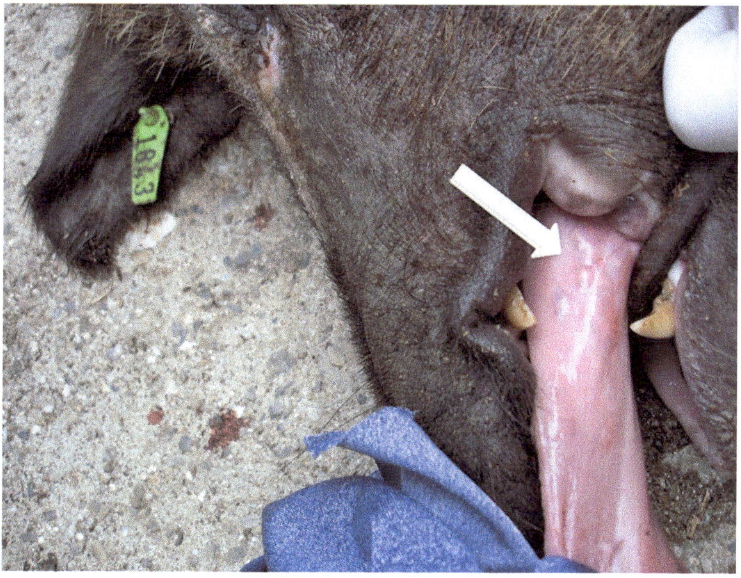

Fig. 10.1 Tongue of infected pig showing nodules of cysticerci (*arrow*)

Fig. 10.2 Pig and man living together. This close relationship is the main responsible for the high endemicity of cysticercosis in rural villages of developing countries (Reproduced with permission from Pathogens & Global Health. Neurocysticercosis special issue, August 2012, cover photo, Oscar H. Del Brutto and Hector H. García, Guest Editors)

Fig. 10.3 Pig corralling in a rural village of coastal Ecuador (Atahualpa). This simple measure prevents pigs to be in contact with human feces and may help to interrupt the life cycle of *Taenia solium*

sewage disposal and may become infected. Confinement of pigs (pig corralling) is a logical measure (Fig. 10.3). It however collides with the requirement of investment from the farmer to feed corralled pigs, something not required if pigs are left to roam and pasture. In survival economies, farmers will be unlikely to comply with sustained corralling because of these economic reasons (Cysticercosis Working Group in Peru 1993; García et al. 2007a). Anecdotally, sometimes open latrines are placed in the backyard where pigs are corralled, allowing the cycle to complete.

Sotelo et al. (1986) reported in 1986 that freezing of infected pork will kill cysticerci in a day at −24 °C or in 4 days at −5 °C. Meat processing by gamma radiation at high doses inhibits the ability of cysticerci to evaginate. Cysticerci irradiated at lower doses may evaginate, but when they are fed to hamsters as a tapeworm infection model, they seem to have an impaired development in the hamster gut (Verster et al. 1976).

Vaccinating the human host to prevent intestinal taeniasis infection has been proposed as a potential control measure (Flisser et al. 2010; Leon-Cabrera et al. 2009), and early experimental models in rodents seem to show a degree of protection induced by *Taenia solium* calreticulin (Flisser et al. 2010). So far, no attempts have been made to produce a human cysticercosis vaccine.

10.2 Current Status of *Taenia solium* Control

In an interesting trial in Cameroon, Assana et al. (2010) used two rounds of porcine TSOL18 vaccination at 3, 4, and 7 months of age plus oxfendazole treatment at the time of the second immunization and found no infected animals in the intervention group 5 months after the last immunization, compared to 20 % in controls; this trial did not intervene in the human population. In Tumbes, Peru, a large-scale control program has been implemented using repeated rounds of human mass treatment with niclosamide, pig treatment with oxfendazole, and pig vaccination with the TSOL18 vaccine, with apparently focal elimination of transmission (Mahanty and García 2010). In the Tumbes program, coproantigen detection was widely applied to detect treatment failures after mass treatment. In the settings of poorly effective regimes (effectiveness of 5 mg/kg of praziquantel in mass campaigns has been reported to be as low as 67 %), a field-applicable coproantigen test could complement treatment efficacy. Coproantigen levels become negative as soon as a few days after successful taeniasis treatment (Bustos et al. 2012).

10.3 Monitoring of Interventions

Monitoring the effects of an intervention requires the selection of practical, feasible, and accurate indicators. Monitoring human taeniasis is impractical due to its low prevalence. Monitoring human neurocysticercosis is hampered by the fact that most cases would be asymptomatic, a significant proportion of them will be seronegative, and brain image is not applicable for this purpose. In the long term, however, changes in the proportion of symptomatic neurocysticercosis cases could be surveyed in people with epilepsy. The prevalence of epilepsy secondary to neurocysticercosis was reduced from 37 to 14 % in Salamá, Honduras, after 8 years of the implementation of a multi-intervention control program for cysticercosis (Medina et al. 2011). This indicator is likely not be sensitive to detect the effect of the intervention in the initial few years, but, on the other hand, it may provide a solid outcome to demonstrate the impact of cysticercosis control in public health.

The evident target for surveillance is the proportion of infected pigs, although no ideal indicator yet exists. Sentinel pigs were proposed and tested in Peru, and they do reflect infection pressure but are not practical because of costs and other logistic reasons (Gonzalez et al. 1994). Porcine infection could be monitored by tongue examination, serology, or necropsy. Tongue examination has the major drawback that it will only pick up heavily infected animals and thus will likely be poorly sensitive for mildly infected pigs particularly after systematic intervention is enacted (Gonzalez et al. 1990; Onah and Chiejina 1995). Serological testing for specific antibodies collides with very high seroprevalence rates in pigs from endemic regions, and many of these animals will likely remain antibody positive for a long period (García et al. 2003). Conversely, antigen testing has the problem of cross reactions with Taenia hydatigena, a very common cestode, and a high detection threshold, likely over a few tens of cysts in the entire pig. Carcass surveillance at slaughterhouses has been discussed lines above and would likely share the very low-sensitivity problems of tongue and antigen testing (Cysticercosis Working Group in Peru 1993; Gemmell et al. 1983). Community participation in reporting infected pigs would likely be needed to survey transmission. This should be designed in a way that it is not associated with losses to farmers in order to keep their cooperation. Improvements in the economy of small farmers could be a great incentive for their cooperation with control interventions. Significant financial gains for the pig farmers following a health and pig-management education intervention aimed to decrease porcine cysticercosis were calculated in Tanzania (Ngowi et al. 2010).

10.4 Control Versus Elimination

Arguments in favor of control and morbidity reduction as a more attainable goal compared to elimination have been published (Engels et al. 2003). Given the very high biotic potential of *Taenia solium*, it is quite possible that control effects will be overcome by the infection pressure soon after interventions are stopped, and transmission will tend to return to pre intervention levels (García et al. 2006). On the other hand, after focal elimination is attained, reintroduction from neighboring endemic regions is an obvious and immediate risk and should be considered for the expansion of intervention programs.

10.5 Foreseeable Future

Confirmation of the efficacy of the proposed control strategies is required as a proof of concept, to later optimize them in terms of easier availability of intervention and monitoring tools without risking the efficacy of control. Drug availability should be improved, the efficacy of taenicidal regimes should be evaluated and optimized, and improvements in the vaccine strategy such as a single-dose, oral vaccine should be listed in the unfinished control agenda. Another interesting possibility is whether control could be performed departing from focal high-risk spots. O'Neal et al. (2012) reported clustering of tapeworm carriers around tongue-positive pigs in an endemic community in Northern Peru and suggested that tongue-positive pigs could be used to define areas for targeted screening or presumptive treatment for taeniasis.

References

Alexander A, John KR, Jayaraman T et al (2011) Economic implications of three strategies for the control of taeniasis. Trop Med Int Health 16(11):1410–1416

Allan JC, Avila G, Garcia Noval J, Flisser A, Craig PS (1990) Immunodiagnosis of taeniasis by coproantigen detection. Parasitology 101(Pt 3):473–477

Allan JC, Velasquez-Tohom M, Fletes C, Torres-Alvarez R, Lopez-Virula G, Yurrita P, Soto de Alfaro H, Rivera A, Garcia-Noval J (1997) Mass chemotherapy for intestinal Taenia solium infection: effect on prevalence in humans and pigs. Trans R Soc Trop Med Hyg 91:595–598

Assana E, Kyngdon CT, Gauci CG et al (2010) Elimination of *Taenia solium* transmission to pigs in a field trial of the TSOL18 vaccine in Cameroon. Int J Parasitol 40(5):515–519

Bern C, García HH, Evans C et al (1999) Magnitude of the disease burden from neurocysticercosis in a developing country. Clin Infect Dis 29(5):1203–1209

Bhattarai R, Budke CM, Carabin H et al (2012) Estimating the non-monetary burden of neurocysticercosis in Mexico. PLoS Med 6(2):1–10

Bustos JA, Rodriguez S, Jimenez JA et al (2012) Detection of *Taenia solium* taeniasis coproantigen is an early indicator of treatment failure for taeniasis. Clin Vaccine Immunol 19(4): 570–573

Croker C, Redelings M, Reporter R, Sorvillo F, Mascola L, Wilkins P (2012) The impact of neurocysticercosis in California: a review of hospitalized cases. PLoS Med 6(1):1–6

Cruz M, Davis A, Dixon H, Pawlowski ZS, Proano J (1989) Operational studies on the control of *Taenia solium* taeniasis/cysticercosis in Ecuador. Bull World Health Organ 67(4):401–407

Cysticercosis Working Group in Peru (1993) The marketing of cysticercotic pigs in the Sierra of Peru. Bull World Health Organ 71(2):223–228

Del Brutto OH, Santibanez R, Idrovo L et al (2005) Epilepsy and neurocysticercosis in Atahualpa: a door-to-door survey in rural coastal Ecuador. Epilepsia 46(4):583–587

Díaz Camacho SP, Candii Ruiz A, Suate Peraza V, Vezueta Ramos ML, Felix Medina M, Lozano R, Willms K (1991) Epidemiologic study and control of Taenia solium infections with praziquantel in a rural village of Mexico. Am J Trop Med Hyg 45:522–531

Engels D, Meslin FJ, Prilipko LL, Saraceno B, Savioli L (2003) Neurocysticercosis: reducing the burden of disease is an even more attainable goal. BMJ 326:511–512

Flisser A, Gonzalez D, Plancarte A et al (1990a) Praziquantel treatment of brain and muscle porcine *Taenia solium* cysticercosis. 2. Immunological and cytogenetic studies. Parasitol Res 76(7):640–642

Flisser A, Gonzalez D, Shkurovich M et al (1990b) Praziquantel treatment of porcine brain and muscle *Taenia solium* cysticercosis. 1. Radiological, physiological and histopathological studies. Parasitol Res 76(3):263–269

Flisser A, Madrazo I, Plancarte A et al (1993) Neurological symptoms in occult neurocysticercosis after single taeniacidal dose of praziquantel. Lancet 342(8873):748

Flisser A, Gauci CG, Zoli A et al (2004) Induction of protection against porcine cysticercosis by vaccination with recombinant oncosphere antigens. Infect Immun 72(9):5292–5297

Flisser A, Avila G, Maravilla P et al (2010) *Taenia solium*: current understanding of laboratory animal models of taeniosis. Parasitology 137(3):347–357

García HH, Gilman RH, Gonzalez AE et al (2003) Hyperendemic human and porcine *Taenia solium* infection in Peru. Am J Trop Med Hyg 68(3):268–275

García HH, Gonzalez AE, Gilman RH et al (2006) Combined human and porcine mass chemotherapy for the control of *T. solium*. Am J Trop Med Hyg 74(5):850–855

García HH, Gonzalez AE, Del Brutto OH et al (2007a) Strategies for the elimination of taeniasis/cysticercosis. J Neurol Sci 262(1–2):153–157

García HH, Gonzalez I, Mija L (2007b) Neurocysticercosis uncovered by single-dose albendazole. N Engl J Med 356(12):1277–1278

Gemmell M, Matyas Z, Pawlowsky Z, Soulsby EJL (1983) Guidelines for surveillance and control of taeniasis/cysticercosis, vol 83.49. World Health Organization, Geneva

Gonzalez AE (1997) Evaluation of a control programme for *Taenia solium* targeting human and porcine health. PhD thesis, University of Reading

Gonzalez AE, Cama V, Gilman RH et al (1990) Prevalence and comparison of serologic assays, necropsy, and tongue examination for the diagnosis of porcine cysticercosis in Peru. Am J Trop Med Hyg 43(2):194–199

Gonzalez AE, Gilman R, García HH et al (1994) Use of sentinel pigs to monitor environmental *Taenia solium* contamination. The Cysticercosis Working Group in Peru (CWG). Am J Trop Med Hyg 51(6):847–850

Gonzalez AE, García HH, Gilman RH et al (1995) Treatment of porcine cysticercosis with albendazole. Am J Trop Med Hyg 53(5):571–574

Gonzalez AE, Gauci CG, Barber D et al (2005) Vaccination of pigs to control human neurocysticercosis. Am J Trop Med Hyg 72(6):837–839

Guo YJ, Sun SH, Zhang Y et al (2004) Protection of pigs against *Taenia solium* cysticercosis using recombinant antigen or in combination with DNA vaccine. Vaccine 22(29–30):3841–3847

Guo A, Jin Z, Zheng Y et al (2007) Induction of protection against porcine cysticercosis in growing pigs by DNA vaccination. Vaccine 25(1):170–175

Huerta M, de Aluja AS, Fragoso G et al (2002) Synthetic peptide vaccine against *Taenia solium* pig cysticercosis: successful vaccination in a controlled field trial in rural Mexico. Vaccine 20(1–2):262–266

International Task Force for Disease Eradication (1993) Recommendations of the International Task Force for Disease Eradication. MMWR Morb Mortal Wkly Rep 42(RR-16):1–38

Johnson RB (1986) Potential hazard of mass praziquantel use. Am J Med 80(6):A88

Keilbach NM, de Aluja AS, Sarti-Gutierrez E (1989) A programme to control taeniasis-cysticercosis (*T. solium*): experiences in a Mexican village. Acta Leiden 57(2):181–189

Leon-Cabrera S, Cruz-Rivera M, Mendlovic F et al (2009) Standardization of an experimental model of human taeniosis for oral vaccination. Methods 49(4):346–350

Mahanty S, García HH (2010) Cysticercosis and neurocysticercosis as pathogens affecting the nervous system. Prog Neurobiol 91(2):172–184

Medina MT, Aguilar-Estrada RL, Alvarez A et al (2011) Reduction in rate of epilepsy from neurocysticercosis by community interventions: the Salama, Honduras study. Epilepsia 52(6): 1177–1185

Molinari JL, Soto R, Tato P et al (1993) Immunization against porcine cysticercosis in an endemic area in Mexico: a field and laboratory study. Am J Trop Med Hyg 49(4):502–512

Montano SM, Villaran MV, Ylquimiche L et al (2005) Neurocysticercosis: association between seizures, serology, and brain CT in rural Peru. Neurology 65(2):229–233

Nascimento E, Costa JO, Guimaraes MP, Tavares CA (1995) Effective immune protection of pigs against cysticercosis. Vet Immunol Immunopathol 45(1–2):127–137

Ndimubanzi PC, Carabin H, Budke CM et al (2010) A systematic review of the frequency of neurocyticercosis with a focus on people with epilepsy. PLoS Negl Trop Dis 4(11):e870

Newton CR, García HH (2012) Epilepsy in poor regions of the world. Lancet 380(9848): 1193–1201

Ngowi HA, Carabin H, Kassuku AA, Mlozi MR, Mlangwa JE, Willingham AL 3rd (2008) A health-education intervention trial to reduce porcine cysticercosis in Mbulu District, Tanzania. Prev Vet Med 85(1–2):52–67

Ngowi HA, Kassuku AA, Carabin H et al (2010) Spatial clustering of porcine cysticercosis in Mbulu district, northern Tanzania. PLoS Negl Trop Dis 4(4):e652

O'Neal SE, Moyano LM, Ayvar V et al (2012) Geographic correlation between tapeworm carriers and heavily infected cysticercotic pigs. PLoS Negl Trop Dis 6(12):e1953

Onah DN, Chiejina SN (1995) *Taenia solium* cysticercosis and human taeniasis in the Nsukka area of Enugu State, Nigeria. Ann Trop Med Parasitol 89(4):399–407

Pawlowski Z, Allan J, Sarti E (2005) Control of *Taenia solium* taeniasis/cysticercosis: from research towards implementation. Int J Parasitol 35(11–12):1221–1232

Peniche-Cardena A, Dominguez-Alpizar JL, Sima-Alvarez R et al (2002) Chemotherapy of porcine cysticercosis with albendazole sulphoxide. Vet Parasitol 108(1):63–73

Plancarte A, Flisser A, Gauci CG, Lightowlers MW (1999) Vaccination against *Taenia solium* cysticercosis in pigs using native and recombinant oncosphere antigens. Int J Parasitol 29(4): 643–647

Pondja A, Neves L, Mlangwa J et al (2012) Use of oxfendazole to control porcine cysticercosis in a high-endemic area of Mozambique. PLoS Med 6(5):1–5

Praet N, Speybroeck N, Manzanedo R et al (2009) The disease burden of *Taenia solium* cysticercosis in Cameroon. PLoS Negl Trop Dis 3(3):e406

Ramos-Zuniga R, Perez-Gomez R, Jauregui-Huerta F et al (2013) Incidental consequences of anthelmintic treatment in the central nervous system. World Neurosurg 79(1):149–153

Sarti E, Flisser A, Schantz PM et al (1997) Development and evaluation of a health education intervention against *Taenia solium* in a rural community in Mexico. Am J Trop Med Hyg 56(2): 127–132

Sarti E, Schantz PM, Avila G, Ambrosio J, Medina-Santillan R, Flisser A (2000) Mass treatment against human taeniasis for the control of cysticercosis: a population-based intervention study. Trans R Soc Trop Med Hyg 94(1):85–89

Sotelo J, Rosas N, Palencia G (1986) Freezing of infested pork muscle kills cysticerci. JAMA 256(7):893–894

Tellez-Giron E, Ramos MC, Montante M (1981) Effect of flubendazole on cysticercus cellulosae in pigs. Am J Trop Med Hyg 30(1):135–138

Torgerson PR, Macpherson CN (2011) The socioeconomic burden of parasitic zoonoses: global trends. Vet Parasitol 182(1):79–95

Torres JR, Noya O, de Noya BA, Mondolfi A (1988) Seizures and praziquantel. A case report. Rev Inst Med Trop Sao Paulo 30(6):433–436

Verastegui M, Gilman RH, Gonzalez AE et al (2002) *Taenia solium* oncosphere antigen induce immunity in pigs against experimental cysticercosis. Vet Parasitol 108:49–62

Verster A, Du Plessis TA, Van Den Heever LW (1976) The effect of gamma radiation on the cysticerci of *Taenia solium*. Onderstepoort J Vet Res 43(1):23–26

Wang QM, Sun SH, Hu ZL, Wu D, Wang ZC (2003) Immune response and protection elicited by DNA immunisation against Taenia cysticercosis. Vaccine 21(15):1672–1680

Wu W, Qian X, Huang Y, Hong Q (2012) A review of the control of clonorchiasis sinensis and *Taenia solium* taeniasis/cysticercosis in China. Parasitol Res 111(5):1879–1884. doi: 10.1007/s00436-012-3152-y

Index

O.H. Del Brutto, H.H. García, *Cysticercosis of the Human Nervous System*,
DOI 10.1007/978-3-642-39022-7, © Springer-Verlag Berlin Heidelberg 2014

MIX
Papier aus verantwortungsvollen Quellen
Paper from responsible sources
FSC® C105338

If you have any concerns about our products,
you can contact us on
ProductSafety@springernature.com

In case Publisher is established outside the EU,
the EU authorized representative is:
Springer Nature Customer Service Center GmbH
Europaplatz 3, 69115 Heidelberg, Germany

Printed by Libri Plureos GmbH
in Hamburg, Germany